# TRANSNATIONAL RESEARCH IN TECHNICAL COMMUNICATION

SUNY series, Studies in Technical Communication

---

Miles A. Kimball and Charles H. Sides, editors

# TRANSNATIONAL RESEARCH IN TECHNICAL COMMUNICATION

## STORIES, REALITIES, AND REFLECTIONS

EDITED BY

NANCY SMALL AND BERNADETTE LONGO

SUNY
**PRESS**

Published by State University of New York Press, Albany



Printed in the United States of America



For information, contact State University of New York Press, Albany, NY
www.sunypress.edu

**Library of Congress Cataloging-in-Publication Data**

Names: Small, Nancy, editor. | Longo, Bernadette, 1949- editor.
Title: Transnational research in technical communication : stories, realities, and reflections / Nancy Small, Bernadette Longo.
Description: Albany : State University of New York Press, [2022] | Series: SUNY series, studies in technical communication | Includes bibliographical references and index.
Identifiers: LCCN 2021048868 | ISBN 9781438489032 (hardcover : alk. paper) | ISBN 9781438489049 (ebook)
Subjects: LCSH: Communication of technical information—Research—Case studies. | Intercultural communication—Research—Case studies. | Research—Moral and ethical aspects—Case studies. | Research—International cooperation—Case studies
Classification: LCC T10.5 .T73 2022 | DDC 601/.4—dc23/eng/20220210
LC record available at https://lccn.loc.gov/2021048868

10 9 8 7 6 5 4 3 2 1

Storytelling reveals meaning without committing the error of defining it.

—Hannah Arendt

# Contents

# Topic-Driven Contents

**External Impacts**



**Focused Preparation**

**Language and Voice**



**Multinational and Multidisciplinary Projects**



**Expanding Research Spaces**

**Personal Relationships**



**Project Evolution**



**Unfinished Business**

# Introductions and Beginnings

Bernadette Longo and Nancy Small

Welcome! This collection is unusual for scholarship in technical and professional communication (TPC). First, it is about research and knowledge production in TPC, but centers on processes rather than products. That said, it is not a "how to do research" instructional text. Instead, it offers commentary and reflection on real-life projects as they unfolded. Second, rather than requiring authors to concisely establish the rigor of their methods, this collection prompts them to share the realities of how research can be complicated, fraught, and even a failure. Rather than presenting themselves as infallible professionals, we asked authors to be vulnerable, to share their moments of struggle and insight emerging out of their past learning experiences. We wouldn't have invited these busy, competent professionals to tell their stories if we didn't firmly believe in the value of their work and if we didn't see a clear need for deeper engagement with questions concerning how research projects are designed, implemented, and written. The nature of their experiences is surprisingly common, but these moments of new awareness, adaptation, questioning, and uncertainty are almost always written out of TPC scholarship. Reviewers, editors, and readers instead equate "rigor" with a produced "perfection," rather than "rigor" as an ongoing attention to quality, reflection, relationships, and agile strategies. A third feature marking this collection as unusual is that the chapters are grounded in storytelling. We chose this mode of writing because it creates spaces where authors can bring themselves—as planners, decision makers, researchers, community members, collaborators, and humans—into texts where they are normally compelled to make themselves *less* visible and disembodied. The kinds of stories found here typically are muted, revised, or erased in our discipline's scholarship.

We envision these storied case studies being useful for a wide range of audiences. Students in qualitative methods or intercultural communication classes can discuss how working across borders complicates the process of designing and completing a study. Scholars looking to expand into projects bridging transnational, situational (e.g., academia and industry), or disciplinary borders might consider these case studies in terms of planning and adaptation. Chapters might inspire brainstorming about the potential challenges a project might face or the moments where meticulous preplanning may need to be revised along the way. Readers interested in intercultural relations and norms or in ethics might ponder the tensions between USAmerican[1] "standard" practices, such as Institutional Review Board or academic journal expectations for "rigor," and the realities of working in complex and relational spaces. For all readers, we hope the collection triggers reflection over how we relate across all kinds of difference, over the power relations inherent in any collaborative situation, and over the accountabilities and reciprocities interwoven with our relationships, particularly as we engage in knowledge-making endeavors.

We argue that, in support of the health and continued growth of TPC, more stories like these should be shared. We focused on sites of transnational and intercultural research because they are complex, requiring navigation of languages, identities, histories, roles, places, cultures, and systems. We also chose stories about transnational and intercultural research because TPC moves among a variety of spaces with increasing frequency and because historically these kinds of projects risk doing harm even when intentions are good. Ultimately, these authors' narratives intentionally *open up* conversation, make visible the embodied and encultured realities of inquiry, and move us—as TPC students and scholars located in richly diverse situations—toward a shared ethic of transnational and intercultural research. Ongoing reflection regarding our practices must be grounded in such conversations as we are our own most valuable teachers.

The remainder of this introductory chapter lays the scholarly foundation for the book, functioning as the disciplinary grounding and literature

1. The adjective "USAmerican" is used because "American" could be from anywhere in the Americas (e.g., Latin American, South American).

review before turning to the readable, engaging stories ahead. We begin by taking a step back, to the story of how this collection came to be. Next, we introduce key terms and basic challenges of learning the specialized craft of transnational and intercultural research. Then we establish the scholarly influences motivating our story-based style, before turning to suggestions for how to use this collection as a teaching, learning, and reflective tool.

If you enjoy the reflective story-based style of knowledge making to come but don't see yourself or your projects in these pages, then we encourage you to find or make spaces for your own narrative work just as we fought to make space for what you see here. While the upcoming chapters move across a wide variety of locations, from North America to the Middle East to South Asia to Southeast Asia and into the African continent, the majority of the voices you'll hear are those of Western-trained scholars. To truly develop a robust transnational and intercultural research effort and ethic in TPC, the conversation must grow to include more perspectives and should continue to complicate our notions of "borders" and interculturalness. Following in the footsteps of our academic predecessors, we offer this collection not only as a contribution to the field but also—and perhaps most importantly—as encouragement for others to tell their stories, too.

## Genesis

We begin by sharing some of our own transformational moments that would eventually orient us toward this project.

In 1994, I (Bernadette) had finished my class work and was facing the research phase of my doctoral work at Rensselaer Polytechnic Institute. When I entered the program two years earlier, I thought I would do my dissertation research on people writing in workplaces. After all, I had been a contract technical writer in the medical and agricultural fields for over ten years before deciding to earn my PhD, and at that time I was most comfortable in workplaces rather than archives. But my interests had changed as I learned more about theoretical underpinnings of technical writing practices, and I was no longer sure what I wanted to study as my own scholarly contribution to the field. So I made appointments with a number of faculty members in my department and asked them about their approaches to research. The one thing I knew at that point was that I wanted my own research to test the boundaries of the scientific paradigm as it applied to humanistic questions. How unscientific could my research be and still be considered valid by colleagues in my field?

My exploration brought me to the office of Alan Nadel, who worked in literary and film studies. He introduced me to French poststructural theories, and I knew that I had found the pathway to my own research goals: I would apply literary theory to technical texts. Dr. Nadel became my dissertation director; he generously supported and challenged my work as I launched into an extended cultural study of technical writing that became *Spurious Coin: A History of Science, Management, and Technical Writing* (2000). The reception of this work encouraged me to continue down the path of humanistic, cultural approaches to understanding the work we do as technical communicators—a path that happily has brought me to this collection.

I (Nancy) was in the midst of my PhD coursework at Texas Tech University. After almost twenty years as a lecturer at Texas A&M University's flagship in College Station, my family and I were invited to expatriate to the international branch campus, Texas A&M at Qatar. With generous support of Qatar Foundation's professional development funding, I finally had the chance to pursue a doctorate. After my first two semesters of drinking from a fire hose of rhetorical theories and after a rigorous introduction to qualitative methods, I signed up for a summer class called Feminist Rhetorics. The professor, Dr. Amanda Booher, encouraged us to explore our larger research interests through critical lenses. Together, we read transformative writings such as Jacqueline Jones Royster's "When the First Voice You Hear Is Not Your Own," Gloria Anzaldua's *Borderlands/ La Frontera*, Gayatri Spivak's "Can the Subaltern Speak?," and scholarship about transnational/global feminism. At that time, my nascent plan for a dissertation project had been centering around Qatari women. While the US news media portrayed women in Arab Muslim nations as oppressed, I had been witnessing a much more nuanced and complicated situation both at work—where women made up 50% of our engineering students—and in daily life. I wanted to understand those gender dynamics and perhaps be a voice in contrast to the negative stereotypes.

For my final project in Dr. Booher's class, I delved into readings about feminism in Arab, Arab American, and Muslim communities, and developed a literature review titled, "For Further Exploration: Framing Contextual Issues in Studying Qatari Women." I look back on that title and its colonialist gaze with deep chagrin, but through that review, I began to realize the questionable (and naive) motivations propelling my interests. My final Fem Rhets class presentation posed questions about the homogenizing nature of the Western perspective, unpacked the inapplicability of Western binaries to Middle Eastern contexts, and explained the rejection of Western feminism in Arab communities. I learned the word *feminist* can carry a stench of colonialism, and that Arab and Arab American women were plenty able to speak for themselves. As I wrote, lessons from my qualitative methods class were ringing in my ears, and I imagined what it would be like to ask my students and other Qatari women to share their stories and perceptions with me, a USAmerican outsider. I thought about how I would have to rely on them to teach me about gender perceptions and dynamics in their communities and personal lives, and that I would be listening to their stories in preparation to analyze and critique them. Just *imagining* it felt awkward, in the kind of way where your subconscious is sending you signals or perhaps even yelling desperately, "This is not a good idea!"

Through that project and more encounters I describe in this collection's final chapter, I talked myself out of my original dissertation plans. I was not the right person to be "studying Qatari women" (so much cringing). Instead, the deep reflection I participated in during Dr. Booher's class redirected me to look inward, at my own community of white USAmerican expatriate women working in Qatar, asking them to share the stories they told about their lives in the region. That project has been revised quite a bit into my own book, *A Rhetoric of Becoming: USAmerican Women in Qatar* (forthcoming). These interwoven experiences—alerting me to the ethical precarity of intercultural study and learning about everyday lives via storytelling—laid the foundation for how I position myself as an inquirer and critical thinker, and those experiences were the genesis of the search inspiring me to propose the seedling of a research ethic also presented in the last chapter of this collection.

Now that I've told my story, I admit I cheated a bit. Bernadette shared her story with me first, in one of our video chats during the COVID-19 outbreak of spring 2020. Before hearing her story, I would have written something different, perhaps about witnessing ethically questionable behavior in a research project, but in response to what Bernadette shared, I reactively reflected, going back further in time, to recall an earlier transformative moment that shaped my values as a scholar. Bernadette's storytelling influenced my own origin story's emergence. In other words, I wouldn't have been able to articulate my deeper motivation had it not been for listening to Bernadette. But that's the thing about sharing stories.

Hearing stories from others inspires us to reflect over and sometimes remember (and re-member) our own. Story sharing shapes and reshapes our relationships with each other, with our disciplines, with our practices, and with our own memories and histories. And now I've just written in this paragraph a story about how a story came to be. As Thomas King teaches us in *The Truth about Stories* (2003), it's turtles all the way down and it's up to us—the listeners, the readers—to decide what to do with the stories we receive.

## Key Terms and Challenges

Working across borders is foundational to technical and professional communication. The discipline's focus, demonstrated in both workplace and scholarly endeavors, is the translation and adaptation of processes and information. As early as 1989, Charles Sides argued in *Technical and Business Communication: Bibliographic Studies for Teachers and Corporate Trainers* (1989) that studies in technical communication drew theoretically from fields as dispersed as studies in communication, reading, psycholinguistics, and human factors. A technical communicator's role was to know the available means of effective communication in any particular situation (to paraphrase Aristotle's definition of rhetoric) and to translate information from one context to another. The role of rhetoric in understanding technical communication practices was actively debated in these early studies, however, as Carolyn Miller articulated in her 1979 article in *College English*, "A Humanistic Rationale for Technical Writing." In it, Miller introduces the idea that communication itself works with other elements in the rhetorical situation to create a context for communication and understanding among people, an assertion that Charles Bazerman and James Paradis expanded on in their 1991 collection *Textual Dynamics of the Professions*. Technical communication scholars since the 1990s—notably Jennifer Slack, David Miller, and Jeffrey Doak in their article in the Suggested Reading list below—have explored implications for power dynamics and the possibility of meaning-making if we think of technical communicators as translators, neutral conduits, or active participants in creating context and meaning in rhetorical situations.

Fulfilling the role of agent in any power-laden communication situation presents challenges, yet in situations where different cultures interact, things become even more complex, interesting, and potentially fraught. Border-crossing endeavors, where TPC scholars and practitioners collaborate with teams to accomplish goals in the making and sharing of knowledge, occur in a wide range of ways. Nancy's "Localize, Adapt, Reflect: A Review of Recent Research in Transnational and Intercultural TPC" (2022) explores compelling variation of our projects, particularly with a focus on transnational and intercultural research with human participants. By "transnational," we refer to TPC projects spanning spatial divides. Although traditional definitions focus on geopolitical borders, the work in this collection can be engaged in projects that are regional or that do not easily correspond to specific nation-states. Movements may be virtual or physical, but are typically complicated by differing time zones, business practices, legal standards, and other systems that require learning, adapting, and developing new relationships. By "intercultural," we refer to working across differences of language, behaviors, beliefs, norms, and traditions. Transnationality and interculturalism overlap in complex ways. Most transnational projects are also intercultural, but intercultural projects need not require transnational movement. Workplace cultures, disciplinary cultures, and local community cultures require us to engage in cross-cultural practices even within our own organizations and locales.

Complicating any conversation about transcultural and intercultural work is the fact that such activities are often rooted in outdated notions of culture, a term that Raymond Williams described in *Keywords: A Vocabulary of Culture and Society* (1976/2015) as "one of the two or three most complicated words in the English language" (p. 49). Williams went on to argue that the idea of "culture" is defined differently in different discourse communities, but could generally be thought of as "a complex of senses . . . about the relations between general human development and a particular way of life, and between both and the works and practices of art and intelligence" (p. 91). In TPC, these "works and practices" predominantly shape knowledge into a scientific way of knowing the world, as Bernadette has argued in "An Approach for Applying Cultural Study Theory to Technical Writing Research" (1998). So it is not surprising that TPC practitioners, researchers, and reviewers consider nonscientific (e.g., story-based) ways of knowing the world to be illegitimate for formal research studies.

Instead of addressing the complexity of culture as a "relation" and "a particular way of life," scholars often approach it via a process of creating catalogs and taxonomies, such as the five dimensions first proposed by Geert Hofstede in his 1980 study of cultural values among IBM employees, *Culture's Consequences: Comparing Values, Behaviors, Institutions and Organizations Across Nations* (2001). These dimensions are power distance, individualism, masculinity, uncertainty avoidance, and long-term orientation. Although this early study extended interpersonal communication concepts in social science, behavioral psychology, and international business, its empirical approach codified the notion of people as units for study. In an updated version of this work, *Cultures and Organizations: Software of the Mind* (2010), Hofstede and his coauthors expand their applications of the original five dimensions, building on this baseline for interpersonal relations: "Human history is composed of wars between cultural groups" (p. 382). Transnational, intercultural interactions may have the potential for warlike relationships, but those of us who enter into these relationships can also keep in mind the consequences of cultural war articulated by Walter Benjamin, himself a casualty of fascist wartime violence. In *Illuminations* (1968), he wrote, "Whoever has emerged victorious participates to this day in the triumphal procession in which the present rulers step over those who are lying prostrate" (p. 256). The authors in this collection tell their stories of situations that could have resulted in domination, but in which we sought better strategies for collaboration instead. We reflect on these moments, as well as our sometimes-imperfect reactions to them, as we seek to compose relationships that complicate or even resist notions of cultural dimensions, objectivity, and conflict in hopes of generative cross-cultural collaborations.

Sometimes these generative efforts go against many tides. Even as we wrestle with learning and enacting better frameworks, disciplinary practices anchor us in restrictive tradition. In academic fields influenced by science and social science, the standard expectations for reporting research via the Introduction, Methods, Results, and Discussion (IMRaD) genre further reinforces de Certeau's idea of objectivity that works to erase transnational and intercultural realities by "educating" human experience into science, described in *The Writing of History* (1988). The IMRaD format itself has come to exemplify a Western model of scientific "truth," but how can truths of human experience across cultures be given legitimacy that resides outside traditional Western science? Instead of relying on an objective measure of truth or scientific validity, the validity of a narrative could be measured by whether "the research prompts further discourse as potential objects of future research. This further discourse would indicate that researchers and other interested parties considered the findings relevant for continued conversations" (see Longo, 1998, p. 64). Using this expanded idea of validity, a narrative could be a valid research report if it prompts ongoing interest in the object of inquiry and findings of the study. In "Humanizing Computer History" (2018), Bernadette argued that stories of people's lives can "bring to light the cultural, political, and economic contexts that influenced the subject's words, actions, and reactions. By immersing the story within its localized context, a [writer] can create a researched story that puts naturalized or objectified objects of inquiry . . . into cultural contests for power and legitimacy" (p. 9). Both Bernadette and Nancy have experienced the diminishment of our lived experiences exerted by the pressure to publish our transnational research in a standard scientific genre. We believe the knowledge silenced by this pressure has value for our own lives, as well as for the lives of our collaborators and colleagues. Some truths about human relations cannot fit into Western science or an IMRaD format.

Beyond complicated definitions, outdated frameworks, and restrictive genres, the challenges of working in transnational and intercultural spaces extend into the publication process. For example, such work should be carefully contextualized according to its particular locations, but when we have provided this literal grounding in manuscripts submitted for peer review, feedback indicated that reviewers often did not want to invest their attention in that substantive information. They wanted to "get to the news" of the studies instead. Based on our own experiences developing a keener sense of relational accountability for our research sites, we felt compelled to reflect on how a project was adapted in response to its location and to the needs of participants with whom we formed relationships. But again, reviewers did not want to know about changes to our plan, the limitations of our Institutional Review Board's requirements, or the ultimate inadequacy of our original research questions or interview protocol. They wanted "traditional" rigor, a plan well-conceived and methodically followed, and specific outcomes qualifying in terms of Haswell's (2005) triumvirate of being reliable, aggregable, and data-driven scholarship. When we looked for advice about preparing for, conducting, and communicating the experiences and outcomes of transnational and intercultural projects, we found a few authors who addressed these challenges head on, but mostly only found hints and breadcrumbs of reflection. Graduate courses and communities of practice prepared us for the concepts and methods of TPC but not for the realities and complications of fieldwork. We devised alternatives—side articles, footnotes, other publication paths—but ultimately each came to admit that we were working counter to central disciplining forces of TPC.

The primary exigency for this collection is to continue making space for a wider range of legitimized TPC student and scholarly practices, including sharing of transformative reflections over the rich complexity of transnational and intercultural projects. The upcoming chapters value contextualization, refute binary and prescriptive thinking about culture, and resist IMRaD as the primary genre for communicating the outcomes of a project. We believe that the stories of our research processes are not "backstories" to the "real" stories but that these stories reveal truths of their own. They go hand-in-hand with the product-oriented outcomes of research reports. As Nancy wrote in "(Re)Kindle: On the Value of Storytelling to Technical Communication" (2017), narrative modes of inquiry and communication have been historically marginalized in TPC. We assert that, in order to fully examine and consider our work, we should commit to more ongoing publication and discussion of our processes. Stories, as tellings of specific events, and narratives, as organized sequences of events into broader arcs of experience and sensemaking, are choice modes for descriptive, contemplative processes.

## Genealogies

In this section, we trace key lines of influence that both inspire and legitimize the scholarliness of the upcoming stories. To begin, arguments for the compelling connections between story and knowledge-making are not new. Indigenous communities long ago established the efficacy of story-based practices, and contemporary scholars continue to contribute important work on the use of story in science (e.g., Kimmerer, 2013), cultural rhetorics (e.g., Brooks, 2006; Cruikshank, 1998; Erdrich, 2003; King, 2003), education (e.g., King, Gubele, & Anderson, 2015), rhetoric and writing studies (Haas, 2007; Powell, 2004; Riley-Mukavetz, 2020), research (e.g., Archibald, Lee-Morgan, & De Santolo, 2019; Wilson, 2008; Windchief & San Pedro, 2019), and more. Individually and collectively, these scholars demonstrate the power of narrative for discussing complex issues located at the intersection of localized symbolic practices and knowledge-making. In other words, the expertise of Indigenous scholars affirms the power of teaching and learning through story-based practices.

Stories effectively convey knowledge because they allow for the existence of "multiple realities"; encourage tellers to set up relationships among the people, places, and events; and entangle both the teller and the receivers in the making of meaning. In short, story-based work allows us to realize that "reality *is* relationships or sets of relationships" (Wilson, 2008, p. 73). Those relationships then invite us to think about the responsibilities we have to ourselves, to our participants, to our research settings, to our readers, to our disciplines, and to our multiple contexts as writers and readers. Positivist-style reporting of a study and its outcomes locks out examination of these relationships, and in doing so, disembodies researchers from their activities.

Story-based scholarship has a small but substantive presence in TPC, a line of predecessors which merits ongoing extension. Here, we specifically acknowledge book-length texts that paved the way for our collection. Breaking ground through its story-based structure, Savage and Sullivan's (2001) *Writing a Professional Life* contributed to ongoing conversations about defining the field itself, by demonstrating the kinds of jobs and concerns TPC professionals shoulder. Alongside it, Bosley's *Global Contexts: Case Studies in International Technical Communication* (2001) emphasized the importance of contextualization as it provided "fictionalized and nonfictionalized scenarios" told from third-person perspectives. Her collection of case studies was designed to uncover "behaviors and patterns of thinking and feeling" as well as "assumptions and presumptions about the cultures from which we each come" and how those affect technical communication projects (p. 3). More recently, Yu and Savage's (2013) *Negotiating Cultural Encounters: Narrating Intercultural Engineering and Technical Communication*, revealed the diversity and complexity of the workplace through first-person stories, encouraging authors to do the kinds of thick description—as well as some interwoven reflection—typically muted or erased in more traditional forms of TPC scholarship. The cases also make excellent teaching tools because they bring workplaces alive for students who have not yet had those rich experiences themselves. While the collections from Savage and Sullivan, Bosley, and Yu and Savage present stories of workplace politics and adaptation influenced by differing expectations and norms, they do not address the exciting and fraught process of setting up a research project designed to move across borders.

Threads of concern over border crossing have been present in TPC for quite a while, too. Two decades ago, Thatcher's "Issues of Validity in Intercultural Professional Communication Research" (2001) questioned the appropriateness of applying traditional notions of research validity to intercultural scholarship. A 2006 special issue of *Technical Communication Quarterly* (*TCQ*), coedited by Scott and Longo, explored the field's "cultural turn" and the associated implications for methods, methodologies, and publication. It was followed soon by another *TCQ* special issue on intercultural communication, this one edited by Ding and Savage (2013). In it, Fraiberg's "Reassembling Technical Communication: A Framework for Studying Multilingual and Multimodal Practices in Global Contexts" (2013) argued that global contexts of study require better contextualization, inherent adaptability, and the space to understand and explain culturally embedded communication. We found related inspiration for this book in two previous edited collections. Thatcher and St. Amant's (2011) *Teaching Intercultural Rhetoric and Technical Communication* explores border crossing through the lens of teaching, while Williams and Pimentel's (2014) *Communicating Race, Ethnicity, and Identity in Technical Communication* focuses on race as a factor in the teaching and production of TPC in the US. Both books amplify the field's continued interest in ethical practices for communicating and collaborating in situations enriched by diversity, so they serve as thought-provoking motivations for this text.

In addition to their foundational work in story-based knowledge making, Indigenous scholars offer important insights into projects situated in intercultural and transnational spaces. Linda Tuhiwai Smith's (1999) *Decolonizing Methodologies: Research and Indigenous Peoples* lays bare the damage done by outsiders dropping in, objectifying the cultures targeted for study, then rushing out, a process we have heard referred to as "smash and grab ethnography" or "parachute research" (see Laura Pigozzi, chapter 6). Smith compels her readers to critically examine their roles—if they are even the appropriate people for a task—and includes an array of examples of Indigenous-centered projects demonstrating a more ethical approach. Shawn Wilson's *Research Is Ceremony: Indigenous Research Methods* (2008) crafts an Indigenous research paradigm from the bottom up, outlining the core elements of ethical project design to include relationality, reciprocity, and respect.

Recent publications continue to teach and model interculturally sensitive methodologies and methods, deepening and broadening our reflections over transnational activities. Windchief and San Pedro's collection *Applying Indigenous Research Methods: Storying with Peoples and Communities* (2019) instructs readers in the craft of "storywork," a highly collaborative, contextual, and multiperspective process sensitive to communicating complexity and helping readers examine compelling, yet historically devalued, methods of knowledge making. *Decolonizing Research: Indigenous Storywork as Methodology* (2019), coedited by Archibald, Lee-Morgan, and De Santolo, demonstrates a broad range of projects where narrative approaches drove rigorous research while centering accountability to the community being studied. These two texts are examples of a chorus of voices offering advice for working across difference via authentic collaboration and projects centered on community interests. Leaders in the Indigenous movement often come from New Zealand, Australia, and Canada, further challenging non-Indigenous USAmerican researchers to position themselves as transnational and intercultural listeners and learners.

While Indigenous scholars often concentrate on projects designed for Indigenous spaces, their advice applies to transnational and intercultural activities in general. That said, don't mistake our respect and amplification of their work as a claim that this current collection performs "decolonizing work." As Itchuaqiyaq and Matheson (2021) remind us, not all work citing decolonizing scholarship actually contributes to tangible decolonizing efforts. Here, we cite and apply the knowledge of Indigenous authors in service of supporting an ongoing shift in disciplinary mind-set, toward valuing stories, contextualization, and reflections. Authors such as Smith and Wilson teach us about the interdependency within which we all live, that everything is built on relationships and their attendant accountability. Respect and reciprocity are ongoing, substantive, and grounded in authentic connections, not transactional exchange of gifts or information. In sum, while scholarship promoting decolonization lays out important paths toward more justice-oriented research and teaching practices, that scholarship offers additional broader lessons in how knowledge is made, what counts as knowledge, and how we might continue pushing back against the boundaries of TPC's disciplinary traditions.

We also find inspiration in the recent social justice turn in TPC, which is concerned with structural inequalities and actions of redress. Angela Haas and Michelle Eble's edited collection, *Key Theoretical Frameworks: Teaching Technical Communication in the Twenty-First Century* (2018), is situated in direct response to globalization and the power imbalances it reinforces. In their introduction, Haas and Eble call on TPC scholars to integrate social justice principles into their projects and to focus on tangible actions that can be taken to disrupt divisive systems and promote accountable intercultural relationships. They affirm the inherent border-crossing motivations and goals of TPC, and remind us of our commitment to a humanistic rationale and that we are positioned to be rhetorical agents of change. Another inspirational text is *Technical Communication after the Social Justice Turn: Building Coalitions for Action* (2019) by Rebecca

Walton, Kristen R. Moore, and Natasha N. Jones. Their introduction highlights TPC's disciplinary problems that a justice-oriented mind-set must commit to address the following: lack of inclusion, lack of commitment, and lack of action. Throughout the book, Walton, Moore, and Jones offer strategies for applying a critical intersectional lens and for engaging in coalition building as twin pillars of more ethical and accountable work in the classroom and in the field. Their framework for strategically contemplating one's role in potentially unjust situations is "the three Ps": positionality, power, and privilege. Reflecting over the relationship among these three interrelated aspects of any setting can prompt awareness regarding damaging hierarchies, dangers of unexamined motivations, potentially fraught collaborations, and systems as well as disciplinary traditions that risk—or even perpetuate—doing harm.

By focusing our collection on sites of inquiry and practice, we dig into one aspect of TPC that is not typically given enough space: the risky realities of our work. As a result, the reflective narratives contained here, by design, examine their authors' positions as researchers and practitioners working in complex intercultural situations. Choices these storytellers confront in their project designs and complications invite purposeful contemplation concerning power, privilege, and agency. By being a conduit for sharing these authors' stories and reflections, we hope to model and perpetuate more socially just efforts in TPC activities.

## Storied Case Studies

We refer to the chapters here as storied "case studies," not in the formal methodical sense of being case studies, but in the spirit of the methodological intent. A case study is meant to be contemplated, viewed through multiple and shifting perspectives, considered in terms of the agents' positionalities, and pondered in its complexity. The style we elicited from the authors is semiformal because we wanted to amplify that you are reading the experiences of "real" people, not overpolished personae. We also wanted the stories of the experiences themselves to be at the forefront rather than being obscured by academese or technical jargon. Names—both of individuals and of organizations—in the case studies are pseudonyms, and we encouraged authors to use detailed and vivid language. Finally, we asked authors to offer their personal reflections in the first person, to share with you what they have learned for themselves but not to tell you what to think. We believe these cases are complex and compelling enough to merit in-depth discussion, and we want you to form your own thoughts and conclusions regarding what emerges.

No two transnational or intercultural endeavors will be alike, so these cases cannot serve as practical models for you to follow. However, each offers its own perspectives on how these projects are designed, adapted, and facilitated. In lieu of a traditional and linear "chapter preview" section, we offer a thematic overview of the cases you can look forward to in the upcoming pages.

Thoughtful preparation is one key theme you'll find. "Planning and Pivoting: Archival Work in Botswana and South Africa," by Emily January Petersen opens with an impressive list of ways she prepared for her trip, ending with a note that she knew even those preparations would not be sufficient. Her opening nicely lays out a theme that traces through the rest of the collection: surprise and (sometimes) serendipity. Breeanne Matheson's "Grappling with Globalized Research Ethics: Notes from a Long-Term Qualitative Research Agenda in India," relates and reflects over a project designed to inquire about the work lives of women in technical communication. Along the way, she describes adapting in response to logistical struggles and culture shock. Matheson concludes with a posttrip reflection in which she reconsiders researcher identity, positionality, and relations to the site itself.

Transformation is another theme threaded throughout this collection. Transnational and intercultural projects do more than create new knowledge: they often result in significant growth and change in perspectives. "Lost in Translation: Losing Rigid Research Team Roles in a Field Study in Vietnam," by Sarah Beth Hopton, Rebecca Walton, and Linh Nguyen, is a three-voice story about the eye-opening benefits of teaming made possible by an openness toward learning about a new culture, as well as by the grace of intercultural relationships. Bernadette's "Accidental Tourist in a Narrative World with Technologies: A Story from Katanga Province" takes us through a narrative of ongoing change in which a project goes through several evolutions in an effort to make an authentic difference in the lives of its participants. Through her travels, Bernadette meets people and experiences the Congo in ways that still resonate with her today.

In addition to commonalities, we note some thought-provoking distinctions in this collection. To begin, you may notice some diversity in our authors' positions in their stories. Yvan Yenda Ilunga's chapter 8, "Relearning Your Knowledge: The Loud Silence," is a narrative of an international scholar finding his voice and making his space in the USAmerican academy. His compelling reflections remind us that listening and adapting are skills central to our home-institutional lives as well as to our transnational and intercultural projects. "Syrian Refugee Women's Voices: Research Grounded in Stories Shared over Coffee Respites" is Nabila Hijazi's story of being a long-term Syrian immigrant who, as part of her Western doctorate education, located a research project in her local Syrian refugee community. Her chapter, like Ilunga's, is set in the US, but demonstrates that transnational spaces are found well inside national borders. Bea Amaya's "Across the Divide: Communicating with Company Stakeholders in Papua New Guinea," offers a unique perspective because hers is the only chapter written strictly from a TPC practitioner's point of view. Rather than embarking on a limited visit to PNG, Amaya relocates there to live in a Port Moresby neighborhood, where she develops friendships with the local families and builds new cultural familiarity through her quiet observations. She then takes us along with her on a trip through the PNG highlands where her company's stakeholder meetings taught her about different expressions of community belonging.

Several chapters complicate our original definition of "transnational" as border-crossing. Laura Pigozzi's " 'Nuesta vida en el medio oeste, USA': Listening to Mexican Immigrants," describes a project set inside USAmerican borders but working with both documented and undocumented participants. As a researcher, she inhabits a liminal space of both belonging to the community in which her project is set and having the privilege of citizen status, a doctorate-level graduate education, and socioeconomic security. Pigozzi's chapter narrates the relational and identity work that her borderland project invoked. Kathryn Northcut's "Chemistry Publication Ethics in China and the US: Transdisciplinary Teaming in a Time of Change" works at the complex intersection of transnational *and* transdisciplinary research. Some of her project's transformative moments impact how she understands collaboration as well as research ethics. A final contribution revealing the potential multidimensionality of border-crossing project is "Mingled Threads: A Tapestry of Tales from a Complex Multinational Project" by Rosário Durão, Kyle Mattson, Marta Pacheco Pinto, Joana Moura, Ricardo López-Léon, and Anastasia Parianou. Spanning locations, it demonstrates how a shared set of methods must be enacted differently depending on specific site contexts. All of these chapters disrupt narrow notions of "working across borders," as that work may take place within national borders or via technologies rather than travel.

In closing, Nancy's "Importing Lessons from Qatar: Toward a Research Ethic in Transnational and Intercultural TPC" is purposefully divided into two sections. The first half is Nancy's story of searching out advice for better practices in transnational research during the years she lived as an expatriate in Qatar. The second half then shifts gears and serves as a forward-looking proposal for an ethic of transnational and intercultural research. Like the rest of the book, that ethic is not presented as a set of rules. Instead, it is proposed as a set of guiding principles and accompanying reflective questions. The goal of this second half of the chapter is to provide a point of departure for TPC researchers and practitioners interested in border-crossing projects.

## Recommendations and Gratitude

Our recommendation for using this collection is simple: read, think, respond, extend. The chapters purposefully do not tell you what to think or assume you agree with the author(s). We encourage you, as our reader, to identify the authors' decisions then consider the complexity of when, how, and by whom those decisions were made. Put yourself in the authors' shoes and ponder what you might have done similarly and differently. Honor the vulnerability with which these authors tell their stories as encouragement to do some deep reflection regarding your own positioning, privilege, and power within the systems that you are striving to function. Sometimes, those systems limit or disadvantage people in ways that affect their research processes and what is considered both "possible" and "academically acceptable." This collection is designed to open up rather than shut down conversation, because it is through such dialogue that our transnational and intercultural practices will continue to evolve and improve. As a concluding tangible contribution to that improvement, the last chapter (Nancy's "Importing Lessons from Qatar: Toward a Research Ethic in Transnational and Intercultural TPC") interweaves observations about these stories into a potential framework for more ethically aware practice. That closing section is *not* meant to provide "answers" to the question "how do I design the perfect project?" Instead, it might get us a few steps down the path of "what should I be aware of and prepared to encounter at any stage of a transnational or intercultural project?" In service of that goal, it provides a preliminary set of guiding questions to help you begin to explore your own strategies for accountability. It also calls for more work developing these heuristics and conversations.

To further facilitate reflection, each chapter closes with discussion questions and with a brief annotated list of suggested readings. The questions obviously are just starting points. Some readers may find it useful to read these questions first, to get a sense of what the chapter author(s) anticipated as fruitful points of contemplation. Other readers may choose to intentionally avoid these questions and allow their own points of interest to emerge as they react to the narratives.

The suggested readings were selected for a wide variety of reasons, and chapter authors included brief annotations to indicate their anticipated usefulness. Some of these curated books, chapters, articles, websites, and videos support the scholarly concepts threaded through the chapter, documenting a source that inspired the author(s) as the chapter was being written. Other selections offer more detail on the specific project or location at the heart of the chapter narrative or because they provide deeper consideration of important issues invoked by that chapter's storytelling. We hope these additional readings serve as useful points of departure as you use this collection to advance your own goals.

In closing, we wish to express profound gratitude to our authors for generously sharing their stories, particularly as we asked them to allow us and our readers behind the professional curtain, to help us experience the realities of being a curious human trying to do a good thing and do the right things. We offer our humble thanks to our transnational partners who generously opened their minds, hearts, workplaces, and lives to us. We acknowledge the scholarly lineages that brought us here, the inspirations we've described in this introduction but also the countless other authors, practitioners, colleagues, and friends whose works have enriched our perspectives. Finally, we thank you, our readers, for your curiosity about transnational and intercultural research, your openness to learning through storytelling, and the insights you'll glean from the chapters to come.

To paraphrase Thomas King, from *The Truth about Stories* (2003), please take our stories and do with them what you will. They may influence your planning or mind-set, or maybe they won't. Just don't say you haven't heard them.

## Suggested Reading

Archibald, J. A., Lee-Morgan, J., & De Santolo, J. (Eds.). (2019). *Decolonizing research: Indigenous storywork as methodology*. ZED Books Limited.

This edited collection expands work on the value and application of Indigenous storywork. It comments on and illustrates the goals and effects of story-centered methodologies and demonstrates the use of storytelling in research in a wide range of disciplines and located in a wide range of nations.

Longo, B. (2018, June 28). Humanizing computer history. *Humanities for STEM Symposium Proceedings*. New York University, April 6–7, 2018. https://osf.io/79ra5/

This paper explores and describes attributes of a humanistic research methodology, along with a consideration of concepts of validity relating to this type of research. It also considers how the writing of biographies constitutes a valid humanistic research method, as well as relationships between archival and biographical research functions.

McSweeney, B. (2002). Hofstede's model of national cultural differences and their consequences: A triumph of faith—a failure of analysis. *Human Relations*, *55*(1), 89–118.

This article provides a thorough and compelling critique of Hofstede's cultural dimensions. It unpacks key assumptions underpinning this ubiquitous theory and argues that all aspects of the study are troubled by undeniable flaws.

Slack, J. D., Miller, D. J., and Doak, J. (1993). The technical writer as author: Meaning, power, authority. *Journal of Business and Technical Communication*, *7*(1), 12–36.

The authors explore different roles fulfilled by technical and professional communicators: transmitters, translators, and articulators. Each role has particular implications for how TPC is perceived as a field and practice, and each role contributes to the production and reproduction of systems of power.

Small, N. (2022). Localize, adapt, reflect: A review of recent research in transnational and intercultural TPC. In L. Melonçon & J. Schreiber (Eds.). *Assembling critical components: A framework for sustaining technical and professional communication.* Colorado State University Press.

This literature review breaks down five years of journal publications (2014–2018) to discover categories and qualities of transnational and intercultural research. The results highlight a wide range of border-crossing projects, and based on the small space allotted to a few authors for reflecting over their processes, finds threads of "better practice" for transnational projects involving human participants.

Small, N. (2017). (Re)Kindle: On the value of storytelling to technical communication. *Journal of Technical Writing and Communication, 47*(2), 234–253.

The author traces the troubled history of storytelling in technical and professional communications, starting with a "narrative turn" in the mid-1990s. Small argues for a reawakening to the value of story-based scholarship and offers an example of how "anecdotes" can indeed serve as "data."

Wilson, S. (2008). *Research is ceremony: Indigenous research methods.* Fernwood Publishing.

Wilson situates and explains his framework for Indigenous research, grounded in relational accountability and foregrounding reciprocity and respect. He works in multiple voices and styles to demonstrate different—and equally valid—ways of making knowledge.

Yu, H., & Savage, G. (Eds.). (2013). *Negotiating cultural encounters: Narrating intercultural engineering and technical communication.* John Wiley & Sons.

This collection of narratives delves into the complex cultures and dynamics in a range of technical communication workplaces, spanning from the automotive industry and manufacturing to security and software development. "Culture" is illustrated broadly as differences in national, regional, racial, and ethnic identity; in disciplinary identity, expertise, and job roles; and more.

## Discussion Questions

1. What are your definitions of *culture*, *intercultural*, and *transnational*? How might the narrowness or broadness of your conceptions affect how you approach these cases?

2. Thinking about your own daily life as well as your special projects, where do you cross cultural borders or work in culturally diverse spaces?

3. We described several inspirations for this collection (such as Indigenous scholars and TPC work in social justice). What threads or themes of scholarship inspire you?

4. We described how the IMRaD genre does not allow space for important kinds of contextualization and lived experience. What is your opinion of that argument? Do you find that the genres you use provide or lack space for certain kinds of information, explanation, or analysis that you believe should be included?

5. What are your thoughts about the value of storytelling and narrative as a means of making and communicating knowledge? Do you have particular expectations or hesitations regarding the *ethos* and *logos* of learning via storytelling?

6. What are your expectations about the nature of transnational and intercultural research? In other words, what do you think you'll find in the upcoming stories?

7. How do you imagine using the stories in this book? How do you expect they'll influence your thinking and planning (or not)?

Chapter 1

# Planning and Pivoting

## *Archival Work in Botswana and South Africa*

Emily January Petersen

When a Batswana archivist told me that "The Director" would take a week to approve our request for copies of archival materials, I felt lost. After a year and a half of preparing to visit archives in South Africa and Botswana to research women's historical technical communication, this hiccup was unwanted. I did as much as I could to prepare for this visit, but it was about to be derailed because I hadn't known to send a request for copies of archival materials in writing a few weeks before arriving. Here is what I had done to prepare, in order to avoid delays such as this one.

- I spent weeks crafting a proposal for funding from the Emergent Researcher Award from the Conference on College Composition and Communication (CCCC), which I received nearly six months later.
- I read nine books and 24 research articles, along with my research partner, who read her fair share of books and articles.
- We crafted a 33-page literature-review document that outlined everything we read, noting the important quotes, practices,

and ideas that could inform our research and our positionality while traveling to collect archival documents.

- I submitted a midterm report to the CCCC about the progress of our grant-funded project, which spanned 19 single-spaced pages.
- I corresponded (over the course of a year and a half) with archivists in Botswana and South Africa via email and Facebook Messenger. I asked about their collections, their hours, the rules, and availability to the public. I checked back in every few months to make sure my visit would still be welcomed. Nobody mentioned the need for a written request for copies during these correspondences.
- I worked through booking flights and hotels and plotting our research schedule at archives in South Africa and Botswana with my research partner.
- I met with a colleague from South Africa to talk through practicalities and possible security concerns.
- We played with apps and technologies to figure out the best option for scanning or taking photographs of documents.

In short, I felt as if I had covered everything. All of this is not to complain. It is to say that a lot of preparation, travel, and work went into arriving at the archives. Much thought went into it, and despite all of those preparatory efforts, which had been sincere, I still hadn't prepared enough. We had visited other countries in the Global South before, so my research partner and I knew what it took to plan a transnational research trip. We also knew that we didn't know everything, so much of this planning was out of an abundance of caution, care, and the knowledge that we would learn even more after we arrived.

## University of Cape Town Archives

When May 2019 arrived, we took a 20-hour flight to Cape Town, South Africa, arriving late at night and exhausted. We found our way to our planned homestay, while staying hypervigilant, given that we were unfamiliar with the area and it was the middle of the night. Then we got some much-needed rest. The next morning, we walked up the hill to the University of Cape Town. We found ourselves in awe of the beautiful view of Table Mountain, the ocean, and unfamiliar flora. While we recognized ourselves to be in unknown territory, we were also on a university campus, a place that feels like home to us. We wandered a little, watching the college students laugh and study, until we found the building that housed the archives.

We entered the door to the vestibule after a security guard unlocked it for us. He seemed to expect that we would know what to do, but we didn't. The procedures were unfamiliar, even if posted clearly on the wall. After some back and forth with him about what we could take into the reading room and how much information we needed to provide while signing in, we made our way up the spacious staircase carrying laptops, pencils, a notebook, and our phones.

We spent the next three days in that gorgeous reading room at the top of the stairs. It had two stories open to the middle, with rows of dark wood tables and a raised librarian's area in the middle. It was still and silent with people studying and thinking. When we first approached the librarians, we didn't know what to expect. We came in with a list of ideas based on our website research, but I wanted to pick their brains. One archivist, Augusta, who wasn't sure by memory which collections they had that would fit our criteria (women's organizations or businesses), set us up at a computer and showed me exactly how to look through their offerings based on a quick search she conducted for me. I thanked her and immediately began filling out slips with call numbers to request that the boxes be brought out for inspection. While I did this, my research partner had located and requested the documents from the Federation of South African Women (FSAW), our original target organization for research. She was now sitting at a microfilm machine scrolling through the black-and-white scans zooming past her gaze. We laughed at how this reminded us of our childhood memories of family history research because of our mothers.

The boxes I requested arrived. And arrived. And arrived. I realized that I had requested much more than we could handle in one day, and that the archivists were graciously filling my requests, without complaint. At the end of our ability to focus, they kindly agreed to hold what we had pulled until the next day, when we would resume our search for women's historical technical communication.

We did this for three days, stumbling on many wonderful brochures, flyers, meeting minute records, and histories. We were able to scan everything we needed with our phones, without complaint from the librarians. When it came time to make sure that we used the images responsibly, I spoke with the young archivist, Augusta, who had helped me to locate so many great collections. She handed me a copyright form and explained that I needed to get permission from the copyright holder and then return the form to them. She gave me her email address so that I could ask questions and complete this process after I returned to the United States. A few months later, I corresponded with her several times to complete this process and ask questions about what to do if the copyright holder could not be found or was no longer organized. She walked me through all of the steps.

Overall, we left the University of Cape Town Archives with rich and promising historical materials that would keep us busy analyzing and recovering women's history for years to come. Even with all of the preparation, we still had to negotiate a few items on the ground—library rules, permission to scan, copyright forms, how to use the catalog, requesting access to collections—but all of it went smoothly because of the people who were there to help us.

Several complicated realities made this personal interaction possible. One is the legacy of colonialism, and an important part of our process was recognizing that we brought this white, colonial gaze with us to the country and to our perceptions of the people we interacted with and the documents we read. Acknowledging this positionality, and the history of atrocities associated with colonialism that made it possible for us to speak English with the people there, is an important part of humility in research and recognizing our own biases. We were able to navigate our time easily, without hiring a translator or asking for much help. Second, we had funding for this project. We were awarded a generous grant that allowed us to travel over 9,000 miles to conduct research and see some of the sights. Our means afforded this experience, one that not many can have. Truthfully, such archives would be better analyzed and interpreted by those who have claim to them, either residents of the country, stakeholders within the organizations, or descendants of the people involved. We also realized that archives are inherently political, and what we found there was preserved because somebody in authority decided it was worthy of preservation. We were interlopers, there with good intentions, but that doesn't mean our presence and gaze weren't fraught. Our research is a site of struggle that highlights our privilege, the realities of history and hierarchy, and the continuing forces of colonialism.

Our experience at the University of Cape Town also highlights the success of our preparations. We knew where to find the FSAW collection, which was also available at other repositories. Once we got there and realized it was on microfilm, we knew that we had likely seen most of what was available elsewhere because the materials had been shared between and among libraries. This saved us further trips to archives in Johannesburg and Pretoria that would have overstretched our timeline. Further, our preparatory reading had given us a solid grasp of FSAW's history and context, so when we looked at its papers, we knew what was important to us and what wasn't. We also figured out through trial and error in our first few hours there that the Dropbox app was best for scanning documents, rather than taking photographs. We were prepared for either reality.

## Botswana National Archives and Records Service

Next, we flew to Botswana. We found our way to the archives, which took almost two hours, despite its nearness to our lodgings, as the taxi did not know where it was, the building was hard to find, and most buildings were not marked with addresses. We knew it was located in the general area where we ended up after leaving our taxi behind and walking toward the government area of the city. This took some 30 minutes, which we enjoyed because it allowed us to take in the sights. I noted the friendliness of the people. As we passed them in the streets or on the dirt sidewalks, they would greet us, saying, "Hello," without hesitation. It surprised me, as people in the United States tend to only greet one another if they know each other. I felt both welcomed and like a spectacle. We were lost and I didn't want to bother anybody with a plea for directions or a lack of respect by addressing them if it was unwanted. I wasn't sure how to behave, despite the friendly greetings. After walking around a fenced government building, and realizing that there was no way out of that compound without returning to where we had begun, we entered the foyer to beg for help. The security man at the desk pointed outside and motioned for us to walk around. Well, we had just walked around the building, but perhaps if we left the fenced area, going "around" would help. It did!

We found the building, nondescript, surrounded with plants, and flagged with a small sign in the yard announcing its identity as an archive.

We entered and again found ourselves speaking with a security guard. We wrote our names in the logbook and were directed toward the archival office, down the hall and around the corner. We entered a room crowded with desks and people, lined up behind the desks and busy with phone calls and computer screens. It was warm, very warm, and we waited quietly for a few minutes until we could make eye contact with one of them. A man took our request, that we were there to look at two specific archival collections about women. I added, timidly, "Are there any other collections about women's organizations that you might know about?" We had identified the papers of two women we wanted to peruse based on my Facebook Messenger interactions with whoever responded to their page's requests. There was no official catalog or computer system, so I wanted to raise my desire for more materials with one of these archivists with the hope that there would be more for me to see.

The tall thin man I had addressed turned to a female coworker and directed her to help us. She pushed a large binder across a tall table in front of the desks that took up most of the room and asked me to sign in and write down what we needed. After that, she sent somebody to retrieve the boxes. When the boxes arrived, we were directed into an adjacent room, which had windows that separated it from the hot busy room full of archivists behind the desks. However, we still had on our backpacks, and we were told sternly that we couldn't take them in. I didn't know what to do. The woman who had helped us let us put them in a small closet and reprimanded us for not using a locker before we entered. We knew that the next day would require that we figure out how to secure a locker in the hallway outside of the office.

When we entered the adjacent reading room, again holding our laptops, phones, pencils, and a notebook, the air conditioning hit us with delicious relief. It was so nice in there. We noticed a few other people sitting at the makeshift desks with mismatched chairs. They were all men, who looked up at us for a few moments, and then returned to their reveries with newspapers.

We settled ourselves at separate desks and began opening boxes and folders. We found many interesting pamphlets and notes from Ruth Motsete and Dr. Gaositwe K. T. Chiepe, whose collection kept growing as she was still alive and donating her papers as they accumulated. There were also some professional papers and letters, but nothing too compelling from the perspective of feminist historiography in technical communication. I felt disappointed, but eventually we found a booklet on women in science and the records for a women's finance house. Those seemed to be promising, if not completely clear as to their value in terms of analysis. The papers we looked through did not fit our research agenda and were, quite honestly, sparse. But we nevertheless wanted a few copies.

I exited the air-conditioned paradise to speak with one of the archivists again. I wanted to confirm procedures with her and ask for another collection, based on some of the information she had given me. When I told her that we wanted to scan a few copies on our phones, she looked alarmed. She explained that we could not take any materials with us, copied or scanned, without written permission from the director. I immediately asked some follow-up questions. Who is the director? How do I get his or her permission? Could I speak with this person today?

She smiled as if I had said something silly, which apparently, I had. She explained that it took at least two weeks to get permission from the director and that we would have to submit a printed letter of request, explaining what we wanted and who we were. Well, I knew that I did not have a printer with me and I knew that I did not have two weeks in Botswana. I explained our situation, that we were leaving the country in a few days. I outlined some of the previous correspondence I had undertaken with somebody in the office, and I begged her to find a way for us to get permission sooner. She told me to write a letter on my computer, email it to her, and she would print it out so I could sign it. I would have to pay for the printed copy, however. I told her that this worked fine and she wrote down her email address for me. We got this done and left. We could not do anything else without the director giving approval. We would return the next morning.

Because much of our preparatory reading had been about South Africa, not Botswana, and because much of what we found about women's organizations was not written about Botswana, we headed over to the National Museum. We got there only to learn that it was closed for renovations. I felt as if nothing was going the way I wanted it to, a feeling that must've made its way onto my face, because the woman behind the ticket window said, "I can let you in anyway, just for a few minutes." She walked us through a courtyard where a hut and path were set up to represent the villages of the Batswana. Past this was a large glass room with a tall ceiling. Inside it stood fancy older cars. She unlocked the door and let us in, following us inside and watching us take it in. It looked like a showroom for a car dealership, but the walls were plastered with large photos and printed text about the history of Botswana. We went around the room, reading all of the information and learning that the cars were used by the country's former presidents. After about fifteen minutes, we finished reading and looking; she followed us out, locking it up behind her.

The next morning, we arrived at the national archives again, this time knowing where we were going and that we needed a locker. We signed in, secured a locker and our things, and walked back into the main office, still crowded with desks and busy people. I felt bad that they didn't get to enjoy the air conditioning that was installed in the next room over, just a window between them. We used the binder to request again the materials we needed, and I anxiously asked if we could scan copies.

The woman who helped us the previous day shook her head. There was still no word from the director. I wasn't sure what to do. I asked if we could scan copies and then wait to get permission by email. I promised that we wouldn't use the images or materials unless we received that permission. She thought about it for a moment and agreed, reminding us that we still needed to pay. I thought about the money belt I was wearing under my shirt and the fact that it held plenty of money. I agreed that this was fine.

We returned to the cool room and got to work, scanning as quickly and efficiently as we could. We didn't want her to change her mind, as she seemed skeptical of us. We had assured her that we were university professors, and even offered business cards, but I understood that we were still outsiders and that made everything we did and said somewhat suspect. They did not know us, and I understood. After an hour or so, we counted out the money, returned the boxes, asked after the director again, and left. He or she still had not approved our request.

In transnational research, sometimes the unscheduled days are the most productive. We spent a few days seeing the sights and eating interesting foods, but we also spent time at the hotel going over the materials we had gathered so far. My research partner is a broad thinker, while I focus in on the details. As we organized the documents we collected and talked over what we were seeing, my research partner talked over some of the trends she was noticing and the observations she had made and I wrote them down, bouncing my own ideas off of hers. This process has been vital to our work in the Global South, as it allows us not only to connect with one another and relax, but it leads to insights and perspectives that only come through reflection that can be thoughtful and unhurried. Because these talks and unscheduled days have been so fruitful, we cherish them, embrace them, and always take a computer or notebook wherever we go.

Inspiration might strike while we enjoy a smoothie at the No. 1 Ladies Coffee House! (We found this shop because of my penchant for reading Alexander McCall Smith's series about Mma Ramotswe in Botswana. It had all of the books on floating shelves around the walls.)

## Johannesburg and Pietermaritzburg

After a few days, we packed our bags and went to sleep, ready to wake early for a flight back to South Africa, Johannesburg this time. However, my research partner woke up much earlier than anticipated with the stomach flu. She was very sick, and I suggested that we reschedule our flight and try again another day. But she was determined to get back to South Africa to keep our tight schedule, and because it was my birthday that day, she wanted to make sure that something happened besides sitting at our same hotel in Gaborone. She rallied and we managed to fly to Johannesburg without further incident. We opted for a hotel within the airport compound that was posh and comfortable, and we ended up visiting the spa for some much-needed relaxation and a little birthday celebration for me! Self-care can be an important part of research travel, especially in unfamiliar places where adrenaline or stress can be high.

The following day, we took the opportunity to visit the Apartheid Museum in the city, an experience that taught us much about the context in which we were researching. The visit to that large and sobering museum codified the historical and cultural information I had been reading about, both in preparation for our trip and in the evenings at hotels. The museum laid out the history of apartheid visually and chronologically, giving me a greater sense of where the archival resources we had gathered were located in time and a better understanding of how what I already knew about the history of the country fitted into a larger, more complicated narrative. This unexpected and unplanned activity ended up being one of the most valuable to me during this research trip because it gave me a cohesive sense of the country's relationship with colonialism, humanizing the people I was studying and laying bare the violence and terror experienced by so many.

We were eventually able to rent a car and drive out of Johannesburg after a few days. We used our time in the car to review some literature that we had downloaded while at the hotel; we wanted to start analyzing, thinking, and writing about what we had seen in some of the archives. This time in the car was productive, with my research partner in the driver's seat, while I was set up with my laptop. Again, we bounced ideas off of each other, and I was able to synthesize some of what I was reading from downloaded research articles.

In the archives of Pietermaritzburg, we had limited human interactions. The room was a bit shabby, but large, and the staff were efficient and incurious. We filled out forms, based on the information we had gathered via the internet before we arrived, and they brought us boxes. We ended up poring over book after book of colonial records. We found mention of "native" women, but no official organizations. However, we quickly began keeping track of and scanning these colonial reports of Indigenous women, realizing that it may be the only records of their voices that exist from that period of time. While we did not find exactly what we were looking for in terms of women's organizations, we did find records of how a hierarchical and colonial organization kept track of its "subjects," and such information will prove to be interesting and insightful once we have a chance to analyze it. We realized that those records gave us an opportunity to expand our vision of what types of records give insight into women's voices and experiences, and that technical communication from colonial sources may provide important and troubling insights into the experiences of the marginalized.

## Conclusion

These three archival experiences, while largely about reading documents and turning through pages, were also about flexible planning and having a spirit of openness. We had planned our research focus and trip details meticulously, taking a year and a half to plan and prepare. We had done everything we could to make sure that our experiences went smoothly. However, that is never the case, as we knew from other experiences, so we also planned by having a good attitude. We knew we had to be flexible. We knew that we had to ask questions and perhaps hear the word *no*. We negotiated these encounters as best we could, and ultimately, we had to respect the rules and procedures of the repositories that house the materials we gathered.

We have never gotten permission from the director since that trip over two years ago now, and I'm not sure that we will unless I reach out again. That's part of the process. We have to keep asking and keep following up.

We have to roll with the punches, realizing that we may not be able to find a building easily or that our presence might not be a happy occasion for everybody around us. We must realize our privilege and accept that. We must learn to be told no and accept it graciously. Our gaze of archival records may not be welcome, and that's something I have to be willing to accept. We must also realize that records might be disappointing. The research we collect may not look the way we hoped it would and it may not answer the questions we have. We can look for ways to change our perspectives and ways to accept what we did find and ask what we can make of it. Sometimes even fruitful research trips, in terms of the amount of information gathered, might not lead to amazing insights or published articles. But the spirit of openness and curiosity that prompts any of us to conduct research outside of our comfort zones, outside of our home countries, and outside of our usual research practices is what made us researchers in the first place.

While we learned a lot by engaging in transnational research, the purpose of it in the first place was to make visible marginalized and underrepresented people. Keeping that focus central is most important when it comes to these types of recovery or research attempts. Understanding that the history of technical communication illuminates the power dynamics that shape the field, and due to my own research, I knew that limited scholarship on women's technical communication history outside of Eurowestern contexts exists. The point of our research trip was to address this problem. A broader understanding of technical communication will emerge from international historical research; such information may change practices and understandings of how technical communication was used and emerged in various areas of the world. This important work is only made possible by the people who engaged with and wrote such technical documents, and we must recognize their contributions and cease marginalizing the important roles they have played. An overwhelming amount of feminist-oriented historical research silences and ignores particular groups of people because of its focus on white, middle-class, North Americans. While feminist scholarship is concerned with uncovering what has been forgotten and highlighting the voices of women, who historically have been marginalized, the field has work to do in including and recognizing *all* women's histories.

I hope that our research eventually adds to conversations of legitimacy, globalization, and writing across contexts for instructors and researchers in technical communication. Uncovering feminist historiographies in various global locations expands disciplinary boundaries and allows us to learn from past practices and various contexts that may instruct and inform the practice and teaching of writing. Expanding the context and locations of our histories will change the way we all view our practices and the way we include different types of knowledge as legitimate.

## Acknowledgments

Many thanks to my research partner, Dr. Breeanne Matheson, of Utah Valley University.

## Suggested Readings

Bennett, J., Muholi, Z., & Pereira, C. (2012). *Jacketed women: Qualitative research methodologies on sexualities and gender in Africa*. University of Cape Town Press.

This book is a result of the African Gender Institute's continental research project, Mapping Sexualities, which was to develop a research methodology suited to case studies of gender and contemporary sexual cultures in Ghana, Nigeria, South Africa, and Uganda. The chapters include questions about what it means to research topics that can be taboo in particular contexts.

Chaudhuri, N., Katz, S. J., & Perry, M. E. (Eds.). (2010). *Contesting archives: Finding women in the sources*. University of Illinois Press.

This edited collection outlines the experiences and methodologies of archival researchers in global spaces. The chapters, which comprise a comprehensive and important introduction to feminist archival research, cover methods, intersectionality, power dynamics, lines of inquiry, and personal experiences.

Petersen, E. J. (2017). Feminist historiography as methodology: The absence of international perspectives. *connexions: international professional communication journal*, *5*(2), 1–38.

This article highlights what is missing in the body of feminist historiography research in technical communication: an international perspective, especially from varied viewpoints and contexts. The invisibility of international perspectives in feminist historiographies suggests that there is vital work to be done in reclaiming and documenting the global history of women in technical communication.

Petersen, E. J., & Matheson, B. (2017, August). Following the research internationally: what we learned about communication design and ethics in India. In *Proceedings of the 35th ACM International Conference on the Design of Communication* (pp. 1–6).

From the experiences of conducting international research for the first time, the authors learned that researchers should be prepared to study and work in ways that recontextualize their understanding of the world based on the unique place in which they are working. Researchers should engage in ethical considerations, taking into account their own footprint as they transit, stay, eat, and work in areas of the world less familiar to them.

Ramsey, A. E., Sharer, W. B., L'Eplattenier, B., & Mastrangelo, L. (Eds.). (2009). *Working in the archives: Practical research methods for rhetoric and composition.* Southern Illinois University Press.

This edited collection offers insights into archival research from a rhetoric and composition perspective. The chapters offer step-by-step guidance for historical research, highlight the principles of archival work, and suggest using the archives in teaching writing.

Walton, R., Zraly, M., & Mugengana, J. P. (2015). Values and validity: Navigating messiness in a community-based research project in Rwanda. *Technical Communication Quarterly*, *24*(1), 45–69.

Community-based research in technical communication is well suited to supporting empowerment and developing contextualized understandings, but this research is messy. Fieldwork examples from Rwanda demonstrate the challenges of turning messy constraints into research openings.

## Discussion Questions

1. How might you prepare for your own international research trip, given the preparations listed in this chapter and some of the unexpected experiences that occurred?

2. How could archival research in a global context fit into your current research agenda?
3. What messy or unexpected experiences have you had in conducting research? How did you navigate those or engage in problem solving?
4. How can you compare and contrast the three archival experiences in this chapter? What do you learn from these connections and convergences?
5. What issues of ethics are raised in this chapter?
6. How might you consider your own privilege when conducting transnational research?

## Author Biography

**Dr. Emily January Petersen** is an assistant professor and director of professional and technical writing at Weber State University in Ogden, Utah. Her research focuses on professional identities and organizations from a feminist perspective by examining social media, uncovering archival sources, and conducting interviews. She has conducted field research in the United States, India, South Africa, and Botswana. Her work has appeared in many journals, including *Technical Communication Quarterly*, the *Journal of Business and Technical Communication*, and *Communication Design Quarterly*.

Chapter 2

# Grappling with Globalized Research Ethics

## *Notes from a Long-Term Qualitative Research Agenda in India*

Breeanne Matheson

Early in my graduate program, I read work from well-known scholars who were working on important globalized projects, which suggested to me that such research, aiming to expand the horizons of the field, was imperative. I was convinced that such work was the most important work I could learn to do. Mentors and friends confirmed that such work was being highly valued in the field; it seemed like an important way to build a name for myself as a scholar.

I was delighted when, in my third year, my graduate school colleague, Dr. Emily January Petersen, invited me to participate in a research project set in India. I jumped at the chance to learn from professionals in a major global technology hub and dove into the task of reading what I could about technical communication in India and its relationship to technical and professional communication (TPC) in the United States, where I live and work.

During her dissertation work on women working in TPC in the United States, Petersen found that many of her participants had colleagues living and working in India. Little research existed about their knowledge and experiences, and we believed that they would have something significant to teach the field. It was in an effort to center the experiences of Indian women that our project was born. Still, despite my experience related to working and traveling abroad, neither Dr. Petersen nor I had any relevant connections to the field in India, upping both the difficulty and the stakes of this endeavor.

Our research partnership was formed simply by identifying an organization that was relevant to our research question (the Society for Technical Communication India chapter), locating the appropriate individuals, and reaching out by email to ask if they were interested in forming a research partnership. Their affirmative answer was the first major step in making our project a reality.

It is a central part of this story to acknowledge that Dr. Petersen and I are both white women, now living and working as assistant professors in suburban, commuter institutions in the USA. Though I hadn't yet recognized it at the time, this positionality afforded me an inordinate amount of privilege that brought about opportunity to work internationally in the first place, despite the fact that I was inordinately underqualified to do so. Those include, but are not limited to, the visibility of scholars who look (white) like me doing transnational work, easy access to institutions of higher education, a graduate stipend and tuition waiver, experience working with international teams from my former career, and access to other researchers with experience in transnational study. Although this project began when I was a graduate student making wages that sat near the poverty line, I had generated enough points on my credit card to fund the flight that allowed me to participate. This resource points to socioeconomic privilege I had enjoyed as a tech worker before I returned to graduate school, which made the entire endeavor of my first transnational project possible. These privilege factors enabled me to learn transnational research but perhaps also emboldened me to attempt a project I lacked the skills for as well.

We set the development and execution of our project on an artificially tight time schedule, to accommodate our academic teaching schedules and our pending plans to go on the job market. This project, we were told, would make us among very few junior scholars who were doing this kind of work. We began working through the details in late March 2016, and set out travel that same July. Given the time required to complete research design, collaborate with the Institutional Review Board (IRB), and plan the logistics of such a research trip, time was tight.

We thought we had done all the right things. We read thoroughly about issues facing women working in TPC and other technology jobs and designed study questions. We underwent review by the IRB and by senior faculty at our home institution and asked individual community members in India to review and provide feedback on our methods. We also held Skype calls with community members on the ground in India who were able to answer some of our questions about the places which we planned to visit and work. Perhaps as a function of our privilege, everyone we enlisted to help us with our project was supportive about our plans. We felt ready.

Having traveled extensively both personally and professionally, I imagined that my culture shock would be minimal. However, upon arrival in Chennai, I felt unmoored by the number of logistical errors we had made right out of the gate. I had a difficult time orienting myself to strategies for safely navigating transportation and lodging, especially having arrived in the middle of the night after two long travel days. We were exhausted upon arrival. We knew that one of our community partnership members, Charvi, would be there to greet us, but we didn't know where we were supposed to meet her. Charvi was not yet there, and we found a spot in the unevenly paved parking lot, crowded with cars and scooters and people, to wait. We kept vigilant for Charvi and when she arrived an hour later, she loaded us into her car and expertly drove us through the streets of Chennai, finally delivering us in front of the homestay we had booked. As we settled in to the homestay, where we had believed we would have two bedrooms, and a sitting area, we found just one bedroom and one bed. Perplexed, we surveyed the room. The lone bed was large enough to fit us both, and we pulled out our travel sheets (much like sleeping bags) to ensure some semblance of personal space. In the morning, cramped and unsure about the unusual lodging in which we found ourselves, we packed up and relocated to a hotel. I rapidly began to doubt my ability to carry out even basic practicalities such as locating suitable lodging and navigating transportation, much less to complete a major research study.

My sense of self-reliance (and frankly, overconfidence) with regards to our practicalities hindered us at several intervals. Before we even left home, Charvi had offered to assist us with booking lodging that would be appropriate for our work. Not wanting to burden her with unnecessary labor, we asked her for some suggestions and I took care of the task myself. However, a cultural gap, likely a result of minor differences between British and American English led us to misunderstand her recommendations. As a result, I booked the accommodations in the apartment where we spent the first night. Not only was that apartment uncomfortable, but it lacked the appropriate facilities for the needs of our project, such as a sitting area for conducting interviews. As a result, we spent much of our first few days on the ground trying to locate more appropriate lodging for our work. Our community partner graciously helped us navigate our missteps and assisted us in relocating. Still, these early logistical difficulties undermined our sense of confidence and used up valuable time that we might have otherwise used to get our bearings before beginning our work.

Another factor we failed to consider carefully enough was our timing once on the ground. Our plans to work in three cities in three weeks made sense on paper, but in practice, our timing was impossibly rushed. Neither of us had ever been to India, but somehow, we had imagined that we would be ready to begin working after only a day on the ground. Charvi had planned interviews for us, and we kept to her schedule, grateful for the steady stream of work that was the key to our project's success. We were meeting so many smart and amazing women, and they were teaching us so much about the culture, workplaces, their lives, and how to interview. One night, we sat down to begin Skype calls with the women who were speaking to us from Hyderabad, and we began to get sleepy. My eyes were drooping and Petersen started to zone out. We could barely focus on reading the research questions in front of us, let alone making sense of our interviewees' answers. We did our best to power through the interview, but we also knew that the interviewee could tell we seemed disengaged. Of course, we were not disinterested, but we were very tired. We found ourselves asking a participant the same question twice in a row, too tired to fully absorb what she was telling us. At the end of that evening, we knew we had been rude, and the next morning we both expressed regret for not being able to give the women the attention they deserved. We knew we could revisit their answers from our audio recording, but we also knew that we could not repair such a bad first impression on our parts. We had failed to consider the intense jet lag we would be facing when we landed in India. Even as the jet lag eventually wore off, our schedule was too tight. To meet our goals, we found ourselves working too long during the day, which we found made us less effective interviewers.

Many elements outside the research project itself also shaped our work and our experience. At times, our values clashed against our work.

For instance, our dependence on bottled water created an ideological challenge as individuals also concerned with the environment. During our trip we encountered intense heat and were drinking a huge number of the bottles of water our hotel provided for our safety, but even a few days in, the pile of bottles that amassed was disturbing. Shortly thereafter, we witnessed a gorgeous river in the heart of Chennai, full of plastic bottles, garbage bags, and other waste and realized that the waste of these bottles was even more pressing due to local limitations around waste disposal and recycling programs. We found ourselves wishing that we had carried a filtered reusable bottle with us. This would have been a more ethically robust solution to meeting our own needs in a global context.

In another challenging event, after almost a week on the ground, I was awakened in the middle of the night by a huge commotion. A troubling domestic incident was transpiring in the hotel room next door. The ongoing noise sent me running to the lobby, barefoot and pajama clad in search of help. Even as I ran to the elevator, I knew that such an incident might easily have transpired anywhere in the world, including at home, where I work as a domestic violence crisis–line volunteer. Even so, I was unprepared for the physical and emotional complexity of being a bystander to a violent event in a new place where I lacked the cultural and procedural awareness of how to respond or intervene. Even once hotel security helped to diffuse the incident and Petersen came downstairs to find out what had happened, I was unable to go back to sleep and spent the morning anxious and disoriented. As with any traumatic incident, that experience also undermined my sense of safety in the days that followed, even though no new threats to our safety had presented themselves. I was entirely unprepared to navigate feelings of trauma while juggling a major research project for the first time. This incident both shaped and hindered our work, demanding we adjust our schedule and our attitude toward our work.

In addition, the culture shock we experienced drew us way away from our original line of inquiry. I was flooded with questions about the cultural and social dynamics we were witnessing, which made our research questions seem smaller and less important than they had originally seemed from my desk in Utah. As we tried to stay focused on our research questions, we continued to learn about the experiences of women working as technical communicators in India. However, our interviews raised a barrage of secondary questions. For instance, we were suddenly very curious about the hired domestic help and mothers-in-law that our study participants reported kept their houses and tended to their children, which made TPC work possible. Our study participants also noted the work and experiences of women working as office janitors and cafeteria workers, invisible work that enables technology workers to work efficiently. These questions about women's work fell outside the scope of our research, but given that we had set out to make the work of women more visible, overlooking hidden labor in technical communication also felt like a painful omission.

In addition to the endless questions we developed beyond the scope of our original project, we rapidly found that our intended study instrument had not been correctly adapted to our population. We had designed our interview questions off of Petersen's earlier work with women working in TPC in the United States with only minor alterations to accommodate some slight cultural differences. It didn't take very many interviews for us to discover that some of our questions were entirely irrelevant and that others used terminology that participants didn't necessarily identify with. While our community partners had checked our work for cultural appropriateness, we had failed to ask about the cultural relevance of our lines of inquiry.

We also found that the concerns and priorities of our participants differed from ours, in part because of complex identity and cultural factors. Further, as we conducted our many interviews, we began to notice the ways that our power and positionality as white, American researchers created difficulties in building trust among our research population. It became clear that our participants were apt to tell us what they thought we wanted to hear or about what would reflect most positively on their communities. Such responses are understandable, given the complex power dynamic that we brought with us as researchers from the United States. For example, at one point, we asked a research participant if there was anything else she wanted us to know about her work as a technical writer. She responded by asking "What exactly are you looking for?" Another responded, "Yeah I wanted to know why the survey and all that you're conducting; do you find coming here useful?" Participants were particularly interested to know how they stacked up, both among their peers in India and those working in the field in the United States. These valid questions began to offer us glimpses into our own power and positionality, which made us appear to be knowledge keepers among our study population.

Another way that our power and positionality manifested itself as participants asked questions was that participants were quick to ask about opportunities for Indian practitioners to obtain education through programs in the United States or wanted us to make introductions to other professionals in the US. Not only were we unprepared with specific information about such opportunities, it rapidly became clear that factors such as the exchange rate between Indian rupees and US dollars would be clear hurdles for such individuals hoping to obtain credentials in the United States.

As a result of these exchanges, the circumstances that allowed us to conduct the research became apparent, revealing a site of injustice in a globalized field. Though we were aware that it was our responsibility to address inequities in the field as we worked, we were, at that time, unprepared with any solutions to address such difficulties. This was, in part due to the fact that though our privilege in positionality appeared high in areas of race and class, I was a graduate student, and Petersen was a visiting professor. Our status as precarious workers inside the field left us with significant limitations. It took us several years and a transition into more permanent positions before we began thinking about tactile solutions toward addressing such inequalities, and longer still to contribute to the solution. In the five years that have followed, we've been deeply engaged in bringing our students into the conversation, developing a course material for Indian practitioners, and participating in the STC India community. Still, these efforts remain limited by our distance, our other professional responsibilities, and our outsider status.

One of the best moves we made as researchers was to enlist the support of a community partner organization. Such partnerships are a vital part of conducting ethical and responsible research. In our case, we worked closely with the STC-India chapter. Both their leadership and community members offered us a wealth of localized knowledge, community introductions, guidance for our study design, and unmatched assistance and hospitality once we reached India. Their support was critical to our project and kept us from failing due to inexperience.

For example, some of our community partners made arrangements for us to visit their places of work where we were allowed to work and conduct interviews. We spent one day in the company of two young women who showed us around their office in Pune. They picked us up at our hotel and drove us through the city. We joined their commute, and chatted about everyday concerns and some of the scenes we saw outside. When we arrived at their place of work, the women situated us in a conference room, shared tea snacks with us, and helped us navigate our interview schedule for the day. We realized just how much we were learning in this moment, about attitudes, class, cultural connections, and belief. Such access would have been difficult or impossible for us to secure independently; the trust and credibility of our individual hosts allowed us to have deeper connections to the place in which we were working and strengthened our ability to localize our work. They also helped to generate enthusiasm among potential participants that made our recruitment process much easier.

Chatting with participants before and after interviews also helped provide us cultural context as we went. In addition to learning about their experiences as multifaceted professionals, they told us about their home lives, their backgrounds, and their goals, which sometimes led to explanations about cultural family practices, social class, religious backgrounds, and historical events that helped us contextualize our data and improve our semistructured interview process.

## Shaping a Long-Term Research Identity

Because of the rich connections we built as we worked during that research project, we rapidly realized that our project could not reasonably be limited to a single visit. First, upon our return home, we encountered many additional questions or inconsistencies in our data. Our community partners were gracious in helping us fill in gaps in our research and in providing member checks about the conclusions we drew. We made several phone calls, exchanged emails, and eventually made plans to return to present what we had learned at the STC India's national conference. There, we met many new people and renewed connections with those we had met on our first trip.

Over time, we began to see our research partners as our peers inside the field and understood that we also had an ethical obligation to return to the community to share the results of the knowledge they generously shared with us. As a result, we made plans to return to the STC India annual conference the following year where we presented the results of our study and conducted an additional questionnaire to triangulate the research we'd already conducted and to extend our research beyond the concerns of individuals identifying as women. We later sent the results of our follow-up study back to our participants by publishing them in the *Indus* newsletter, a widely read publication that serves to connect Indian technical communicators to each other and to the field. Still, even our plans for regular visits and community engagement left us feeling as though we were so geographically and culturally distanced from the site of our work that our research would always remain shaped by our outsider status.

We also began to understand the quandary we were faced with during that first visit; our participants were concerned about their ability to obtain additional formalized education in TPC in ways that were accessible to a range of professionals and affordable in the context of complex economic factors. We eventually moved to mobilize our students in a knowledge sharing project, developing free educational resources for Indian practitioners that could be accessed remotely and asynchronously to account for the wide variety of needs present in that population. Our report on the details of that project were published in the *IEEE Transactions on Professional Communication*. Though such a project represented a complex collaborative pedagogical strategy, such a move allowed us to begin to engage with (but not solve) injustices inside the field as they relate to individuals outside the Eurowestern frame often prioritized in TPC research. Further, such endeavors inherently prioritize Eurowestern systems of education and power by positioning US-based students as expert and India-based professionals as learners. As with everything in a globalized frame, our every move was fraught with issues of power and position.

Our ongoing work reinforced something we knew to be true from the TPC literature but didn't fully understand until we were deep in the thick of it: transnational research can easily become a colonial possession of knowledge that doesn't belong to you. In the absence of clear remedies to the injustices baked into the system of academic inquiry, I am aiming to make such injustices visible, to consider my contributions to them openly, and to make moves toward justice, pressing against academia's colonial structure whenever possible.

## Upon Reflection: Some Insights from COVID Travel Restrictions

We originally planned to carry on our engagement with this community via regular visits and research projects and by way of service-learning with our students for years to come. These plans were halted in 2020 at the onset of the globalized pandemic around COVID-19. As infections spread around the globe, grounding planes and shuttering professional gatherings, I knew almost instantly that this long-term research partnership was going to end up on indefinite pause.

A year of working from and staying primarily inside my own home shifted my research plans toward thinking about projects closer to home. In particular, I began thinking and writing about the field's use of decolonial methods, along with Dr. Cana Uluak Itchuaqiyaq. As part of our work (see the Suggested Readings), we conducted a corpus analysis of decolonial methodologies in the field. We noted that many scholars, myself included, used decolonial frameworks as a metaphor for social justice–related topics.

Thinking deeply about Tuck and Yang's 2012 article "Decolonization Is Not a Metaphor," in *Indigeneity, Education & Society*, pressed me to think deeply not only about how frameworks are used but about who should use them; the limitations white, EuroWestern scholars bring to the table; and about how such scholarship easily replicates the violent structures of the settler colonialism that has shaped much of the history of the Global South. This work has led me to question not only how scholars like me should work globally, but whether we should do so at all.

It was during this period of study and contemplation that I began to take more seriously that my research projects in India might have been more ethically conducted by individuals who had long-standing roots in the STC India community. At a minimum, we should have spent much more time in a place before writing a study instrument and provided considerably more time between the conception of a project and the design of the study. Such preemptive work would have allowed us to begin to identify ways in which our positionality led us to overlook important cultural elements to a project. Further, we should have included a local primary investigator or simply have turned the project over to community members entirely.

It's easy to see how we got here: academic structures currently constrain such long-term projects and encourage scholars to forge ahead with quick turnarounds in order to produce first-authored texts. These constraints were particularly relevant to us as early career researchers facing deadlines related to graduate school funding, job market forces, and tenure clocks. Such structures compelled us to rush every stage of the research process, limiting our efforts to assess whether we were correctly positioned to do such research, and restricting opportunities to engage deeply with a community before endeavoring to begin working preemptively. It is essential for departments and programs housing TPC graduate students and faculty to reconsider the existing parameters of measuring successful research if they wish to promote ethical, deep research as a priority in the globalized field. Further, good mentorship should advise students carefully and frankly, offering detailed foresight about the pitfalls students endeavoring in globalized projects are likely to encounter.

With globalized travel in an ongoing state of ambiguity, it seems likely that our continued plans to carry on this research project will remain on hold in the short term. In the meantime, I'll be deep at work considering what the future of this project should hold or whether it has reached an endpoint.

## Acknowledgments

Thanks to Dr. Emily January Petersen for her partnership in this research endeavor and her support of this manuscript, to Dr. Cana Itchuaqiyaq for her brilliant scholarly partnership, and to the Society for Technical Communication India Chapter for its hospitality, generosity, and brilliant knowledge sharing.

## Suggested Readings

Itchuaqiyaq, C. U., & Matheson, B. (2021). Decolonizing decoloniality: Considering the (mis)use of decolonial frameworks in TPC scholarship. *Communication Design Quarterly Review*, *9*(1), 20–31.

As noted in the text above, work on this piece was key in helping me conceptualize the complications and limitations present in doing global and cross-cultural work, especially as a white, US-based scholar. I suggest this piece as a way of thinking about how frameworks often used when studying multiply marginal populations can be easily appropriated or misused by scholars who are immersed in the Eurowestern nature of academia that is often complicit in the marginalization underrepresented groups.

Matheson, B., & Petersen, E. J. (2020). Engaging U.S. students in culturally aware content creation and interactive technology design through service learning. *IEEE Transactions on Professional Communication, 63*(2), 188–200.

This article reports on a globalized service learning project conducted as a follow-up to the project described in this chapter. This classroom case study provides a model for what sustained, reciprocal research could look like along with the limitations of such an endeavor.

Spivak, G. C. (1999). *A critique of postcolonial reason.* Boston, MA: Harvard University Press.

This text was foundational to our study design, helping us think critically about the complexities and ethical challenges embedded in transnational research. We found this critique and others, by scholars of color globally to be a vital part of conducting responsible global research.

Tuck, E., & Yang, K. W. (2012). Decolonization is not a metaphor. *Decolonization: Indigeneity, Education & Society, 1*(1), 1–40.

As noted in the text, this article provides an important perspective on thinking about notions of decoloniality that have often been co-opted to stand in for social justice related topics. This is key reading for individuals in the field hoping to engage in global research, especially when originating from the US and Europe.

Thorp, L. (2006). *The pull of the earth: Participatory ethnography in the school garden* (Vol. 7). Rowman Altamira.

This book provides an in-depth look at how research projects might look when they endeavor to be long term and community engaged. The author found it to be particularly helpful when designing community-based research for the first time.

Walton, R., Moore, K., & Jones, N. (2019). *Technical communication after the social justice turn: Building coalitions for action.* Routledge.

This text provides vital thinking about how technical communication can operate in more socially just ways. In the context of research design, this text offers ways of helping researchers think about the power dynamics they bring to a research situation and offers suggestions about how to handle complexities of privilege and positionality in responsible, ethical ways.

## Discussion Questions

1. What do you see as the primary ethical issues arising from the positionality of the project in this chapter? What responsibilities does the researcher have in responding to them?
2. Thinking about the ways the author has framed her positionality and limitations with regard to her work, how would you frame your own positionality and limitations in relation to your own research projects?
3. The author mentions several ways in which she and her research partner attempted to mitigate their own privilege; can you think of other strategies they might have used?
4. Thinking about notions of decolonial research, what issues can you identify with the way the field often encourages Eurowestern scholars to work in the Global South? What obligations do you think Eurowestern scholars have to engage in decolonial scholarship efforts?

## Author Biography

**Dr. Breeanne Matheson** is an assistant professor at Utah Valley University. She has a growing amount of experience conducting international field research in the Global South, and interests in teaching technical communication and rhetoric through community-centered projects. Her most recent research seeks to understand the technical communication strategies employed by activists in South Africa to fight racial inequality and discrimination against women.

Chapter 3

# Lost in Translation

## *Losing Rigid Research Team Roles in a Field Study in Vietnam*

Sarah Beth Hopton, Rebecca Walton, and Linh Nguyen

The exigency for our project originated in Sarah Beth's personal experience with the complicated and contested human health consequences of dioxin exposure. Her stepfather, a Vietnam veteran, served as a medivac recon helicopter pilot during the war and was stationed at the Da Nang Air Base, which stockpiled Agent Orange throughout Operation Ranch Hand and until the end of the war. The Air Force base was so contaminated it was the first site selected for environmental remediation in 2012. Her stepfather's exposure to Agent Orange resulted in several health issues, each exacerbating the next, including chloracne, non-Hodgkin's lymphoma, neuropathy, and eventually early-onset Alzheimer's disease, which took his life in 2016. Her sister, his biological daughter, also suffered complicated reproductive issues likely resulting from her father's exposure, as dioxin is particularly effective at disrupting the endocrine and reproductive systems of women. Though her stepfather received a diagnosis of "presumptive exposure," which entitled him to full medical disability, her sister did not, as the children of Agent Orange victims both in the United States and

Vietnam are not extended the same benefits as their parents, even though evidence suggests that dioxins are epigenetic.

Sarah Beth's personal experience watching two people she loved suffer from exposure-related illnesses compelled her to take up the rhetoric of Agent Orange in her dissertation work. She met Dr. Rebecca Walton, who became her research partner, shortly after defending her dissertation, which investigated the complicated rhetorical constructions of Agent Orange. In discussing that project, they were both struck by how "wicked" the problems around communicating the contested science and policy practices of exposure were in both the United States and Vietnam, and they wondered if technical communication might have something important to contribute to those transnational conversations. Rebecca brought years of qualitative research experience to such a project. She was also a prolific writer, editor, and generous collaborator. Sarah Beth brought in-country experience to the project, as she had lived in Vietnam for a year as a teacher; a broad background in the science, policy, literature, and rhetoric of Agent Orange; significant and successful fund-raising experience, and of course, personal experience of the human health consequences of exposure.

Their goal was to conduct field research grounded in feminist and transnational methodologies, which is to say that they wanted their sponsoring organization, the Vietnam Association for Victims of Agent Orange (VAVA), to generate research questions and goals *with* them. Generally however, they knew they wanted to study the communicative practices of an international, humanitarian organization around issues of contested science.

Rebecca went to Vietnam in October of 2015, to connect with several potential partner organizations and reached out to VAVA expressly. As business in Vietnam is highly relational and best conducted in person, these initial contacts were a critical step in establishing their credibility as researchers and allowed their project to quickly move forward that summer. In less than a year, they successfully generated approximately $10,000 for the project by winning various institutional grants from both universities, which would cover their in-country expenses for five weeks. They secured the necessary visas and government permissions and began the Institutional Review Board (IRB) process in spring of 2016.

## Hiring a Translator

To conduct this research, they also needed to hire a translator. They knew that translator experience level, hours of availability, and salary expectations would vary widely. They were strategic and explicit about setting priorities: they wanted to hire someone they "clicked" with, who was invested in the purpose of the research, and who could work flexible hours, including travel. As a trade-off, they determined that professional experience in research or even in translation was not required, so long as the person was a first-language Vietnamese speaker who was also fluent in spoken English. They preferred to hire a translator who lived in Hanoi, where VAVA is headquartered, to facilitate communication between the research team and VAVA before they arrived in the country.

Once the priorities, timeline, and budget were established, they were ready to fill the position. They used all the personal and professional networks at their disposal to spread word of the position: friends who had traveled to Vietnam, colleagues at their universities whose expertise took them to Vietnam, family members, and friends of friends—no one was too far removed from their network to be contacted. To applicants and potential applicants who inquired about the position, they sent a follow-up email identifying themselves, summarizing the research project, describing the job, and presenting the compensation.

This highly networked strategy paid off. A friend of a friend who'd taught English in Vietnam posted about the position on his Facebook page. Linh, who became their translator, did not know him personally but had friended him online to learn more about English and to hear of scholarships and other opportunities he would occasionally post. She sent him her résumé in response to his post about the job opening, and when she didn't hear back promptly, she assumed it was a scam, but he forwarded her résumé to Rebecca and Sarah Beth, and it caught their attention.

Linh's résumé was among the first of several they received, and they were drawn to her largely because of her bio statement, which supplemented the summary of her skills and experience elsewhere in the resume by conveying a sense of what it might be like to work with her: "I'm energetic, always get excited about new things, and cannot wait to express my creativity. I have a quick learning mind yet keep my eyes on details." They followed up with Linh by email to share more details about the job, and Linh's response strengthened their sense that she would be a good fit, saying that she found the research purpose "interesting and meaningful" and wished to contribute to that purpose as an interpreter. They interviewed Linh by Skype to further gauge her language skills, which clinched their decision. Linh not only had the language expertise, but she also had great energy, an enthusiasm for learning about research, and a sensitivity to the research topic. The team was born.

## What We Learned from Each Other

### Learning Professionalism: Rebecca's Unexpected Lesson

During our five weeks in-country, we learned a lot from Linh about what it means to "be professional" in the contexts in which we were working. I was in my 30s, a professor with three graduate degrees who'd conducted research in multiple countries on four continents. I was experienced in conducting interviews in general and conducting interviews through translators to study humanitarian organizations in particular. So you might think I'd know plenty about how to comport myself as a professional. Linh had recently completed her university studies. This was her first job as a professional translator and the first research project she'd worked on, but she had extensive knowledge of professionalism we lacked, specifically as it related to the intercultural nuances of showing respect, being appropriate, and engaging professionally.

This expertise reflects a major premise of transnational mentoring, particularly models based on comentoring: all people have things to learn and things to teach. Traditional mentoring models are predicated on unidirectional learning from the top down. In contrast, comentoring models are intentionally nonhierarchical and reciprocal, with people moving in and out of "mentor" and "mentee" roles. So it was important that we were open to learning from Linh, as she was to learning from us.

For example, we were aware that it was important to present gifts to the organizational leaders we would meet with at the beginning and end of our in-country visit. We thought that gifts we had selected from our university bookstores would be a good choice, since these items were obviously from our local area and represented our employing organizations.

Linh was not impressed with our selections.

She educated us about proper gift giving, helping us craft personalized messages of thanks on the cards we had brought for those purposes and helping us select additional, supplemental gifts so we could send the genuine message of gratitude we intended to convey.

Our research took us to 11 provinces in Northern and Central Vietnam, and we spent a lot of time traveling by car. It was hot in Hanoi and even hotter the farther south we traveled. We had packed what we considered to be appropriate professional dress for embarking on a multihour car ride to visit regional offices of the nonprofit organization: for example, below-the-knee skirts with short-sleeved pullover tops or modest sundresses over leggings with nice-casual sandals. No jeans. Covered shoulders and knees. Everything modest and polished.

Linh was not impressed with our selections.

It took a few days for her to broach the topic of our wardrobe, but eventually she suggested that when we packed for the upcoming flight to Danag, we might be sure to bring professional clothes.

"What do you mean? We're wearing skirts and dresses. Nice clothes."

"No, it's not really professional. You don't even have a collar. See what I'm wearing? My shirt has a collar."

She continued the lesson on what makes clothing professional for the rest of the ride back to our hotel, explaining subtle signals that conveyed professionalism and respect. We jotted a few notes down and we were both surprised to learn that exposed tattoos were not necessarily a deal breaker, whereas lack of collar really downgraded the perceived professionalism of an outfit. That evening, we went shopping, doing our best to follow the rules of professionalism. The next day we displayed our purchases for Linh's approval.

Linh was impressed with our selections. We were learning.

And the learning went both ways. Having worked with translators in research interviews before, I suggested that the three of us review the interview protocol together and that Linh jot a brief glossary in the notebook she used during interviews. This glossary of unusual or technical terms would be useful when translating in the moment, shortcutting the work of finding just the right word or figuring out how to explain a particular concept. We also discussed best practices, such as jotting key words while listening to participants to better remember their words and inform more accurate translations, as well as establishing a signal for requesting that a participant pause to allow Linh to translate what had been said so far.

As the research continued, we learned more about comporting ourselves as professionals to signal our respect for the people generously spending their time with us and sharing deeply personal and intimate stories and details of their lives living with dioxin. We dressed professionally. We presented appropriate gifts and received gifts with genuine gratitude. We were careful to greet people in an appropriate way and appropriate order: depending who was in the room, we greeted in the order of seniority, starting with the first person in the line and going in order. We accepted offers of hospitality, often beginning meetings with tea.

I think it's important to acknowledge that, as cultural outsiders, we *couldn't* enter the country with a contextualized and complete understanding of how to signal our respect for those we were meeting, to include the subtleties of comporting ourselves as professionals within that cultural context. Certainly, we tried to prepare before the trip: meeting with those who had cultural expertise to share with us, reading not only scholarly and historical materials about Agent Orange but also blogs and online advice for visitors to Vietnam. But, as the saying goes, you don't know what you don't know. To plan ahead for learning things we didn't know we needed to learn, it was essential to hire a *local* translator: not just a bilingual person, but a member of the local community who would be willing to serve as a culture broker to help us navigate cultural differences respectfully.

Simply telling someone—someone in a position of less power, someone who is younger than you, someone who is in your employ—to let you know when you do something wrong is not sufficient to cultivate a climate of transnational mentorship. Several things contributed to the success of the feminist mentorship model that set the stage for Linh's cultural guidance. To address this, we wrote into the job description that the translator would serve as a culture broker. In her initial job interview, we explained that we sought a translator who'd really want to be a part of the team, someone who could help us to learn more about Vietnamese culture, who would help us to understand the meaning that people were trying to convey in interviews, and who would help us to convey the respect and gratitude we felt for research participants and the nonprofit organization they represented. In country, we demonstrated respect for Linh's knowledge by planning site visits together among the three of us, by listening for Linh's subtle or indirect signals, and by following up and inviting more direct discussion when the three of us were debriefing and preparing for the next round of interviews.

When the three of us met together for the last time before we flew back to the United States, we had thought hard about what we'd learned from Linh. We would give her bonus pay, acknowledging the excellence of her work, and we would also provide her with detailed letters of recommendation on our respective university letterhead to support her job prospects going forward. These were practical gifts that we felt confident would be appreciated, but we knew they weren't quite enough. So after much discussion, we arranged through the hotel to purchase an enormous arrangement of flowers, recognizing that the size of the arrangement would reflect the level of gratitude. We presented the flowers to Linh while explaining how essential and important her contributions were, how much we enjoyed working with her, and how much we loved and would miss her. We practically glowed with pride when she praised our gift and exclaimed that we had learned: Linh was impressed with our selections.

## Learning to Play: Sarah Beth's Unexpected Lesson

Technical and scientific communication traditionally has been an agonistic, male-dominated field that historically met the needs of the scientific and military industrial complexes and later large institutions and corporations. Traditionally, it is a field that values plain, direct language over creative or descriptive language; rewards logic and critical thinking over emotional appeals or intuition; and tends to organize itself through highly defined sets of rules, documents, or systems that often compartmentalize, standardize, and, ideally, professionalize. Being attentive, exacting, and methodical makes for an excellent technical communicator, but such communication styles don't always travel or translate well across contexts, generations, or genders.

When we traveled to Vietnam to conduct our study, I was a new, untenured professor in my late 30s who had spent nearly two decades in Florida politics. When I went back to school in my mid-30s I spent a good deal of time and energy worrying that I wouldn't be able to catch up or keep up with my peers intellectually, as most had extensive backgrounds in the rhetoric and technical communication scholarship and literature, something I didn't have, even as I had years of professional experience. My natural tendencies toward logical and critical thinking, compounded by graduate school anxieties around inclusion and acceptance, converged to create the belief that in order for my science to be valid, I would need to remain objective, discerning, and emotionally distant. So I approached the relationship with our language translator as one might any other capital transaction: we would compensate her fairly for her time and talent and in exchange she would translate and negotiate on our behalf. Of course, that's not exactly how it happened.

Linh was a recent graduate who, we later learned, had never in her life spent a night sleeping away from her family, yet no matter the social class or person of prestige with which we interacted, Linh carried herself with uncommon self-possession rarely seen in someone so young. She was an incredibly talented translator, but also a crackerjack cultural ambassador, ensuring our dress code violations or gifting gaffes didn't permanently hobble our research goals, and she was also a logistics genius, facilitating complicated schedules across the variable terrain of 11 provinces that required negotiating, booking, or ordering various modes of transportation, accommodation, and food round the clock.

Her work with us was demanding, difficult, and exhausting, and she would often fall asleep on the taxi rides between interview sites. It wasn't simply the work of translation that was stressful, but also, and perhaps especially, the absence of her family and friends, with whom she was incredibly close. As a Western woman known and praised for her objectivity and emotional distance, I did not at first understand—nor was I very understanding of—Linh's need for connection and play. Not only did I think this crossed boundaries that might complicate the manager-contractor relationship, but I was also concerned about objectivity: would getting too close invalidate our research findings somehow?

Part of the unease I felt was the result of my own general discomfort with Southeast Asian "girl culture." I don't use the word *girl* to indicate chronological age or to be dismissive but to denote a youth-oriented, feminine cultural focus, a state of mind, a celebration and amplification of "all things girl," which is prevalent in many Southeast Asian cultures, including Vietnam. Scholars who study girl cultures note that the practice of girl culture offers the girl both agency and oppression, meaning sometimes "girl culture" behaviors seem to conform to mainstream cultural values (like Japanese girls wearing their hair in traditional buns) and at other times these behaviors resist tradition (like dying the bun a wild color and integrating the bun into a "Cosplay" costume). I didn't understand this tension very well because it was a community of practice I had never participated in, and so when Linh wished to Snapchat filter selfies as a way to connect and relate to each other during time off, it felt, well, *strange* and unprofessional.

After several weeks in the loud industrial city of Hanoi, Rebecca and I were looking forward to the slower pace and natural beauty of Hoi An, with its gorgeous rice fields and its beautiful mountains overlooking the South China Sea. We were also looking forward to a few hours of solitude, which we'd been unable to enjoy due to our packed and exhausting schedule. The three of us agreed to meet first thing in the morning for the next day's interviews and then went our separate ways. For Rebecca and me, the evening offered rest that was a long time coming; for Linh it proved to be a very long night.

The next day when we met downstairs for our 12-hour day, Linh was visibly exhausted. When we asked her why she was so tired, she told us that she'd never slept away from home, didn't sleep well and was tired and anxious. In the taxi on the way to the interview site, we asked Linh what would make her feel better, how we could help. She thought about it for a while and then said she needed to "just hang out with us." When we got back into Hoi An, we spent the rest of the day together not as colleagues or professionals but friends, ready to "play" and connect with Linh in ways that were important to her and that *she* defined. We walked the streets of Hoi An shopping for clothes and hats. We took selfies and giggled as we reviewed the roll of silly faces and googly eyes that stared back at us. We drank pastel-colored bubble (boba) tea, skipped the streets with arms interwoven, visited temples, and had our fortunes read.

We'd finally figured out that being present for Linh in ways that made her feel cared for and connected also recharged her, and even enabled her to do the job we'd hired her to do better. By the end of our time in Hoi An, Linh's energy had returned and she was negotiating meals, hotels, and gifts for our hosts and translating complex IRB-required documents and helping us clarify our interview notes with enthusiasm and skill. She had gotten used to sleeping alone, and we had recognized that shooting a few silly-faced selfies didn't disqualify our research findings, but rather enabled them. I still prefer serious to silly, and never quite got used to the taste of bubble tea, but we found an unexpected cure to the demanding and, at times, disquieting work of interviewing victims of Agent Orange exposure in the renegotiated terms of our relationship to Linh, the engagement with girl culture, and the more flexible practices of transnational mentorship, which, surprisingly, better supported our technical work.

## Learning to Enact Integrity: Linh's Unexpected Lesson

When I was first introduced to the translation opportunity with Sarah Beth and Rebecca, I was a young Vietnamese girl who had just finished a degree in International Business from the Vietnam National University. Half of my professors came from the United States, and our curriculum in the International Business program was inspired by that of University of Illinois. I studied Principles of Law, a compulsory course, about the US legal system, earning an A+. The longer I stayed at school, the more I learned about Western ways of working. In addition to attending a Western university, I was also chosen to present my native country of Vietnam to work with people from around the world in international competitions and conferences. All together, these experiences gave me the confidence to apply for the translation position, considering I felt I understood people who lived on the other side of the world map. I approached the translation job with conservative learning expectations, meaning, I expected Sarah Beth and Rebecca would help me expand my language skills, improving my vocabulary and pronunciation. Fortunately, I was wrong. I learned far more than just language.

Before we interviewed anyone, I needed to translate the letter of information and share it with interviewees every single time we conducted interviews. This letter was part of the Institutional Review Board requirements, something I wasn't familiar with. The document began with common clauses like an introduction/purpose of the interview and procedures that we would follow during the interview, but then there was a clause called "Risks," and that's where all the weird experiences began.

The "Risks" section stated clearly that there might be a risk of discomfort to participants. Discussing Agent Orange with Rebecca and Sarah Beth, who came from the very country that manufactured and dispersed it, might feel awkward for participants and they wanted to acknowledge this up front in an effort to reduce the discomfort and learn from participants. Additionally, this section referenced confidentiality. Sarah Beth and Rebecca took extra steps—three layers of protection, in fact—to ensure participant confidentiality and data protections. Additionally, the document stated that participants could stop and even withdraw from the interview anytime, for any reason.

Vietnamese culture is an Eastern culture, which means we have an indirect way of communicating. We try to avoid conflict and disagreement as much as possible. As an example, I was taught that if I disliked someone, even intensely, I should never tell them this to their face, even if they said something that made me feel uncomfortable. We do this because we are afraid of damaging or losing relationships with others, even if they don't have a positive impact in our lives. At the same time, we are a culture of sharing. We share our food and share our private lives with others. Sometimes, our neighbors know our family in more ways than we might imagine. They know where I studied, for example, how much scholarship money I earned to study there, and even how much money I made while working after graduation. Consequently, we don't protect personal identities or value individuation. So, the term *confidential* would only be used in legal paperwork. That such language was used in an interview document was strange to me and to participants, but after some time, I understood that they were offering protections both to participants' identities and also showing concern for their emotions. From this, I learned some lessons, specifically that it was important to respect people's emotions and identities, even if the person doesn't have a sense of individuality or expectation of confidentiality themselves.

I started every conversation with participants by translating each clause and reviewing with participants each clause. Indeed, starting a conversation involving such sensitive matters like Agent Orange victimization is not an easy thing to do when the person who sat in front of you lost their arm in the war, or birthed children with deformed limbs. As I talked through each clause, I saw doubt change into understanding and cooperation. I wondered if it was one of the rare instances where their identity was respected and emotions around Agent Orange, appreciated. As a Vietnamese person, I had never thought about it like this before.

Learning to translate and relate the clauses was simple, but my lessons were as yet incomplete. As I thought about it, I was trying to name what I was seeing the two professors try to do. It wasn't just basic respect, it was also something bigger than that which I couldn't name, but the more I worked with the two professors, the more I realized I was experiencing *integrity*.

The letter of intent, as the document was called, also included a clause called "IRB Approval Statement," with a link to the website and contact point at the two professor's universities in case participants had questions about their rights and wanted to seek redress from people other than Rebecca and Sarah Beth. Rebecca and Sarah Beth signed the document and then requested me to sign as a witness, after which they carefully handed over to interviewees the document with both hands, an important sign of respect in Vietnam.

Although Vietnam is a developing country with a robust economy, our English-speaking population is still a minority, especially uncommon among older generations who were also those most likely to be VAVA representatives and thus participants in our study. If you had taught in Vietnam, you would also know that students rarely raise questions during classes. This is because expressing ones' self publicly or representing ones' self as an individual is not common practice in our socialist republic. As a result, the chances of having anyone question their rights or contact a US university or representative to clarify or invoke rights was at 0% chance of happening. So, why would Sarah Beth and Rebecca need to offer such protections if there was little or no chance they would be used or even needed?

Because this is the meaning of integrity: abiding a set of morals and ethical practices transparently, without compromise or exception. The two professors did not—and would never, I must say—consider the fact that participants wouldn't reach out or question their rights, rights that they did not even know that they had, or didn't care to honor themselves. Rebecca and Sarah Beth presented themselves consistently as we traveled the country together. They never forgot to sign the paper, and reviewed the LOI over and over again, even when participants tried to get them to hurry through or dismiss it. They expressed themselves in a way that was professional and dedicated. Rebecca put her hands on her heart whenever the word *risk* was mentioned, to express empathy and humility. Their hands were on their chest whenever we talked about interviewee rights as a guarantee that they were well-protected. All of these acts were done to help communicate protections and unconditional respect.

Growing up Vietnamese and engaging Eastern cultural practices daily, I thought I knew the best way to work with my own people, but Sarah Beth and Rebecca, from the opposite side of the world, in some cases, knew how to communicate respect and trust better than I did, even as the language with which they spoke was not English, but universally recognized symbols like touching your head and heart and arms to signify wisdom, love, and strength. They also taught me a whole new language that will help me harmonize with anyone in the world, and they expanded my definition of integrity.

## Wrap-up

Recognizing that some aspects of our lessons learned in the earlier sections are context specific, we end by distilling some broader takeaways applicable to our future work, particularly in transnational contexts.

*Balance preparation with openness:* On the one hand, it's important to learn everything we can to prepare for the environment, culture, and stakeholders associated with a particular transnational project. But at the same time, we must acknowledge that there will be gaps—the most dangerous of which are things we don't know that we don't know.

*Expect to teach and to learn:* Regardless of our official title, role, or job description, we must seek out opportunities to learn from our teammates. If we're in more powerful roles, we hold a greater responsibility for opening space, for inviting others' knowledge, and for demonstrating humility.

*Formally encode expectations for mutual learning:* We learned that it's not enough to rely solely on comportment (e.g., "being nice") or vague statements (e.g., "Just let me know if you have anything to add") to encourage mutual learning. We must be intentional and structural in seeking out expertise, especially from team members in positions of lesser authority and especially early in the project. Write expectations into job descriptions (e.g., the "culture broker" aspect of our translator ad). Establish norms that enable multidirectional learning (e.g., planning the day's interview schedule together as a team). Begin as we intend to continue (e.g., before conducting interviews, review the protocol with your translator and ask which questions might need to be revised together, ordered differently, or even cut altogether; this not only improves the interview protocol but establishes the norm of seeking out and valuing a translator's cultural expertise).

*Don't run from discomfort:* Mentoring can be uncomfortable, especially when we are called to "unlearn" things we were pretty confident about. It can be embarrassing to be corrected or instructed (e.g., learning how to dress professionally). It can feel awkward to meet mentees' needs, especially psychosocial needs (e.g., bonding with teammates over selfies and bubble tea). It can seem pointless to engage in completely new practices (e.g., explaining the rights of research participants). But this "unlearning" is not losing ground; it's a valuable enrichment. Sometimes discomfort is a signal that mentorship is working: that we have an opportunity to learn an expected-but-valuable lesson.

## Suggested Readings

Aspen Institute. (August 2011). "Agent Orange/Dioxin History." Retrieved: July 6, 2020. https://www.aspeninstitute.org/programs/agent-orange-in-vietnam-program/agent-orangedioxin-history/

This website offers an overview of Agent Orange history, remediation efforts and the diseases related to dioxin exposure. The website also offers further reading and links to information on the chemical composition of dioxin, Congressional testimony that resulted in the Agent Orange Act of 1991, and other helpful sources one can study in order to understand the complexity of the scientific, rhetorical, and political constructions signified by the term *Agent Orange.*

Cain, M. A. (1994). Mentoring as identity exchange: conflicts and connections. *Feminist Teacher*, *8*(3), 12–118.

This article details the author's personal experiences with mentors, both male and female, and argues that "gifts" of identity are complex, beneficial in some ways but creating unexpected obstacles in others. The author explores the power of the female mentor and suggests that other women need female advocates to succeed as students, retain professional positions, and work toward institutional and social gender equity, work that women, she suggests, are best equipped to do.

Eble, M. F. (2008). Reflections on mentoring. In L.G. Eble & M. F. Eble (Eds.). *Stories of mentoring: Theory & praxis* (pp. 306–312). Parlor Press.

This book offers an overview of the status of mentoring in the field of composition and rhetoric, a sister field to technical communication. Eble and Gaillet use 80 contributive articles to highlight how, why, and in what ways mentoring is an important topic, and they highlight the critical means and methods used to prepare graduate students to meet the demands of professional development, gender, and tenure issues, and the enculturation of new faculty members and administrators.

Gonzales, L. (2018). *Sites of translation: What multilinguals can teach us about digital writing and rhetoric.* University of Michigan Press.

This book would be useful to those planning transnational projects that involve multilingual communication. Particularly if you are a monolingual person preparing to collaborate with interpreters for the first time, this book can help you develop an understanding of the multilayered, nuanced rhetorical work required to translate from one language to another.

Keller, E. J. (2018). *Rhetorical strategies for professional development: Investment mentoring in classrooms and workplaces.* Routledge.

This book is a valuable resource for those who want to more deeply understand mentoring and its relation to teaching and learning. The author investigates mentoring in both workplace and classroom settings, investigating the ways that gender identity plays a role in the experiences of mentoring and being mentored.

Moore, K. R., Meloncon, L., & Sullivan, P. (2017). Mentoring women in technical communication. In *Surviving sexism in academia: Strategies for feminist leadership* (pp. 233–240). Routledge.

This chapter presents the feminist mentoring model employed by the organization Women in Technical Communication, focusing on two key aspects of mentoring: (1) using dialogue to flatten hierarchies and (2) "deploy[ing] affect through listening and recursive orientation" (p. 234). This chapter is a particularly good resource for those seeking mentoring models that oppose systems of oppression such as sexism.

## Discussion Questions

1. Have you ever been in a strong, positive mentoring relationship? What was your role in that relationship: mentor, mentee, both? What made the mentorship successful?
2. Have you ever had a negative mentoring experience? What did you learn from it that will inform your own mentorship practices? How might those practices inform your transnational project?
3. What do you think enabled this team to learn from each other? What qualities, strategies, or other considerations helped team members shift in and out of mentor and mentee roles?
4. The authors list four lessons learned about engaging in transnational projects. Choose one of the four lessons and discuss what it might look like in the specific context of your own project. How might you implement this lesson yourself?
5. Each of the team members in this transnational project learned a different unexpected lesson. How might good mentorship practices set the stage for you to learn unexpected lessons in the course of your own transnational project?

## Author Biographies

**Dr. Sarah Beth Hopton** is an associate professor of technical and professional writing and director of the English Internship program at Appalachian State University in Boone, North Carolina. Her technical writing scholarship focuses on the intersections between social justice, regenerative agriculture, and technology. Her research has appeared in *Technical Communication, Technical Communication Quarterly, Present Tense, Communication Design Quarterly,* and in several edited collections. Her article *All Vietnamese Men Are Brothers,* coauthored with Dr. Rebecca Walton, recently won the 2020 CCCC Best Article Reporting Qualitative or Quantitative Research in Technical or Scientific Communication.

**Dr. Rebecca Walton** is an associate professor of technical communication and rhetoric at Utah State University and the editor in chief of *Technical Communication Quarterly*. Her research interests include social justice in sites of work and qualitative methods for cross-cultural research. Her coauthored work has won multiple national awards, including the 2020 CCCC Best Article Reporting Qualitative or Quantitative Research in Technical or Scientific Communication, 2018 CCCC Best Article on Philosophy or Theory of Technical or Scientific Communication, the 2016 and 2017 Nell Ann Pickett Award, and the 2017 STC Distinguished Article Award.

**Linh Nguyen** is a translator and eCommerce customer manager working for the global consumer products manufacturer Unilever. This is her first coauthored publication. She lives in Ho Chi Minh City, Vietnam.

Chapter 4

# Accidental Tourist in a Narrative World with Technologies

## *A Story from Katanga Province*

Bernadette Longo

"Why don't we give them all iPhones?" asked a graduate student in that semester's Information Design class. The "them" he referred to were small-scale farmers, small businesswomen, and artisanal miners in Katanga Province, Democratic Republic of Congo (DRC). It was spring semester 2010 and the student who asked this question was new to the project; I had been working on it for three years and had to remind myself that there was a time when I might have asked this question, too.

I often taught master's level courses in information design at the University of Minnesota—a course topic that needs a specific focus to make the content relevant. In late 2007, I was looking for a partnership with a community organization as a course focus. I hoped to find a community partner who would be interested in establishing a long-term relationship with my classes to develop and carry out a communication design project that would benefit their nonprofit organization while providing learning opportunities for my students and me. I don't know about my students, but I learned more than I could have imagined.

## Hello? Are You There? Can You Hear Me?

This version of my story begins when I started planning for a spring semester information design class sometime in 2007. For 15 years, I had taught communication and information design via interdisciplinary partnerships and collaborations with community organizations, so I was not coming to this semester's project as a "service learning" novice. I talked with an acquaintance who had started an NGO to help women start businesses in the DRC—in this chapter I'll call it Women in the Congo (WITC)—with US offices in Minnesota and DRC offices in Lubumbashi. This partner and I determined that we would work with my information design graduate students to address organizational communication questions that could benefit WITC, while also providing an opportunity for students to learn course content. So we planned the course.

Over the years, I had experienced political tensions in community-based research between people who use their local knowledge in everyday transactions and professionals, like myself, who might use that knowledge for career advancement. I tried to exchange university resources for local knowledge, but always felt the ethical tension of these exchanges. During spring 2008, my students and I conducted a communication audit for our partner organization and determined that their website needed redesigning. Students focused their efforts on elements of the website that could be more persuasive with more dynamic images and narrative content. At the end of the semester, we made detailed recommendations for redesigning the website and identified a firm that would carry out the redesign and host the site pro bono. Our partner was thrilled with these outcomes and some students continued working as volunteers with WITC after the class ended.

I questioned, though, that women in the DRC were actually benefiting from our design work in Minneapolis. Were these women working "on the ground" actually contributing their needs and ideas through our NGO partner? I asked our partner whether women or WITC staff in the DRC had access to the website we had just worked on and whether it had relevance to their day-to-day operations. The answer was "No." Of course, people in the Congo did not have much access to the internet, even if they did have computers (which was also unlikely). So my afterglow from our semester's work started to fade, even though we had benefited US operations for WITC.

I asked our partner whether women and WITC staff in the DRC owned cell phones. The answer was "Yes, almost everyone in the DRC owns a cell phone." In 2008, this question about mobile device use was not as obvious as it is in 2021. Communication researchers were not thinking so much about delivery of information via mobile phones, especially in the US where these devices were used mainly for voice transmission versus data transmission. It seemed to me that if we wanted to benefit our colleagues in the DRC, we should concentrate on communicating via cell phones. Our WITC partner and I decided to focus on this expansion of our work for the next information design class in spring 2009. I started looking into design and communication issues involving information delivery via cell phones and prepared for the next class.

During spring 2009, the students and I worked with our partner to explore what kinds of information our colleagues in the DRC wanted to communicate to help them conduct their businesses and coordinate their operations. Our partner and I developed four research questions.

> How do students in my class, WITC staff, and WITC entrepreneurs describe their social network?
>
> How do students in my class, WITC staff, and WITC entrepreneurs currently communicate with others in their social network?
>
> For what functions do students in my class, WITC staff, and WITC entrepreneurs currently use their cell phones?
>
> What expectations do students in my class, WITC staff, and WITC entrepreneurs have for using their cell phones to keep in contact with their social network?

From these research questions, we developed interview and survey questions that we could ask both our colleagues in the DRC and students in my information design class. The process of developing these questions already brought up significant differences in worldview between our two groups of collaborators. For example, our students would understand the term *social network* as a discreet part of their lives, perhaps a network based both in face-to-face and technology-mediated experiences. But we had difficulty using that term social network with our DRC colleagues. As our partner explained, our Congolese collaborators would not understand a social network apart from the fabric of their lives. So asking about a social network as something that was separate from their lives as a whole would not make sense to them. We determined that we would ask about their *business circles*, which was a WITC term that designated their 10-woman self-help groups and could be understood as distinct social units. Again I experienced the unsettling feeling of how difficult our dialogue would be if we did not share an understanding of social networks and connections among human beings. I wondered how much common ground we could find and build on. I also wondered about the limits of "science" in a Western tradition as a framework for transnational and transcultural collaborations. It seemed to me that this scientific model of "validity" and "reliability" did not hold when people were attempting to bridge incommensurate worldviews.

With the help of our partner and one of the students from the previous year's class, we set up a Skype call to interview two business women and one WITC office worker in the DRC. Our partner served as an interpreter for our conversation. Although some of our DRC colleagues knew some English, they were more comfortable speaking in Swahili or French. I did not know either language, but could understand very limited French terms. We relied on our partner to interpret our colleagues' responses, although she could not literally translate entire parts of the conversation that went by quickly. Our internet connection was also somewhat fragile and we experienced a number of dropped calls during our 90-minute conversation. We would call back, but between the technology interface, lack of visual cues, and language interpretation, our dialogue was partial at best. Yet without these affordances, I would not have been able to talk with my colleagues in the DRC at all. Again, I wondered how much we could bridge the divide, even with good intentions on all sides.

During the interviews, I asked the following five questions of each of our three DRC colleagues:

> What does your business circle mean to you? How would you describe your business circle to someone who didn't know about WITC?
>
> How do you stay in touch with the people in your business circle? Do you talk with them every day? Do you use your cell phone to stay in touch?
>
> Please tell me how you currently use your cell phone.

> How would you like to use your cell phone?
>
> Please describe how you connect to people in rural areas now and how you would like to improve those connections.

We recorded the conversation and posted it on the internet to share with the students and potentially with our colleagues in the DRC. I doubted that our colleagues in the DRC would be able to access the information, but I felt I had to at least keep trying to strengthen our dialogue in any way I knew how.

We also distributed questionnaires to the students in the class via email and received their answers. We asked these questions, which we tried to align with those used in the DRC interviews:

1. What part of your social interactions is most important for you in your closest social circle? You can choose to describe what you value most about your life and relationships with the people who are close to you, such as your business circle, family, friends, or other social group.
2. How do you currently stay in touch with the people in your closest social circle, such as your business circle, family, friends, or other social group? How do you communicate with the people in your closest social circle?
3. Please let us know how you currently use your cell phone.
4. Please let us know two ways you would like to use your cell phone to do things you do not currently do with your cell phone.

Generally, we found that both groups favored face-to-face communication as their primary contact with people in their social groups. However, the students reported that they probably communicated as much or more with people in their social groups via telephone or other technology-mediated modes, even though they favored face-to-face communication. When these interviews were conducted in 2008, most people in the DRC used simple Nokia-type phones that were not able to connect to the internet. These phones enabled voice and simple text message (SMS) communication. Our Congolese colleagues communicated much more face-to-face, using their cell phones to contact people in their social groups mostly in urgent situations, due to the high cost of cell phone minutes. They reported that they would like to use their cell phones more for business, but would have to be able to pay for the minutes. Our WITC office staff member also said he would like to have a smartphone so he could connect to the internet. In this aspiration, he was more like our student respondents, who also said they would like to use their phones more for internet connections and downloads.

The next step was to explore options for a cell phone-delivered communication system that would support WITC operations in the DRC. Our partner said that one communication problem they had was connecting women in the urban area, where the organization was headquartered, with women in a rural area about 73 kilometers (45.36 miles) away where they were expanding their operations. It was difficult for WITC staff and the women in the rural area to stay in touch with each other through monthly face-to-face visits. Because the women in the rural area had cell phones and could receive SMS text messages with no charge, we determined that we should explore the capabilities of delivering information that would support WITC operations via simple text messages.

When presented with this information design challenge, students in the information design class initially suggested that we find someone to donate smartphones to our colleagues in the DRC, which would provide them with applications to share information and coordinate their business operations. When we explored implications of introducing a new technological device into this cultural context, we determined that this initial solution probably represented a common approach to solving communication problems, no matter what the context. "Apply more technology to the problem" seemed to be a "commonsense" approach to us. But we had not carried out an appropriate design process that included our potential users' participation. For example, our DRC colleagues already told us they had difficulty affording cell phone minutes with their simple phones. How would they afford to use their smartphones? Would their cell phone infrastructure support this type of smartphone? As we posed these and other questions regarding the appropriateness of our first approach, we quickly decided that our ideas about a high-tech "fix" would not actually help our colleagues in the DRC achieve their communication and business goals. And because this solution did not include expertise and information from our DRC colleagues, we had not engaged in a user-centered—not to mention human-centered—process.

From an ethical point of view, we needed to involve our DRC colleagues in design decisions about this communication system to help their businesses. I knew there was a great deal about life in the DRC that I did not understand. I could talk with our partner and get one point of view on the circumstances there. I could research the area online and in publications. But until I went there myself, I knew that I could never feel what life was like for our colleagues on this project. I began talking with our partner about traveling to Katanga Province in January 2010.

## Meeting Colleagues in Katanga

We finished our information design class in spring 2009 having completed some initial exploration of the communication system our partner requested, but also knowing that the more we found out, the more complex this design project became. We found ourselves asking the most basic questions about cell phone use in the DRC because we realized that we did not know anything about the situation there. We had to start with the most elementary questions about how people used cell phones there because we could not make assumptions based on our own experiences in the United States. At the end of the semester, we really had more questions than answers about how to design a mobile phone–delivered communication system to help businesswomen in Katanga.

Planning the trip to Katanga was an exercise in strained dialogue that lasted for most of the summer and all of the fall. Our partner and her family in the DRC were concerned that I would be shocked by the living conditions I would see there and tried to prepare me for daily life in the neighborhood where I would stay. I thought I was a seasoned traveler who had been in many unexpected and unfamiliar situations far from home, so I got myself prepared for the unexpected. Our partner and I had many conversations about this aspect of the trip. As the trip got closer, I also began receiving emails from people in the DRC who were working with our partner to organize the trip. It was clear that I was being perceived as someone with great funds who would pay many hundreds of US dollars for services I might need in the DRC, like a guide or a translator. Our partner went back and forth in lengthy email communication about the fact that I was not wealthy and could not pay large amounts of money for services. I appreciated the fact that our partner could negotiate these elements of the trip and realized that I would not be able to continue this collaboration at all if it was not for our partner's intervention. On my own, I could not negotiate this dialogue across our cultural divide.

I landed in Katanga right after the new year in 2010. Lubumbashi is a large city with 10 million people, but it is not "urban" in the ways a Westerner might recognize. Remnants of a history of brutal colonization coexist with 50 years of troubled postcolonial politics in a big city caught between massive riches of natural resources, exploitation, and day-to-day life. There are many stories I could relate about my visit to Katanga. But I will concentrate on two of them that relate most directly to the text messaging project that is the subject of this case study. The first is about the day we drove about 75 kilometers (46.60 miles) to Kasumbalesa, a rural area at the Zambian border where women affiliated with WITC were doing business. Some farmed; some had small shops that were similar to convenience stores. All twenty women who lived and worked in this rural location had cell phones. I met with representatives from this group of women and enjoyed a meal that they had prepared for my visit. We then sat together and I asked them about their businesses and their use of cell phones, using the same questions I had asked earlier during the Skype interviews. Their description of their cell phone use was similar to that described by our collaborators in that earlier Skype call. I felt like I was getting somewhere with the project because at least I was hearing corroborating description for the assumptions that we were working with from those interviews nine months earlier. In the midst of this unfamiliar situation, it felt good to hear this familiar story.

As our conversation continued, though, women began talking more about how they operate their day-to-day business, and I realized that although we could address some of their business challenges with a cell phone system, they had bigger problems that we could not touch. For example, women who did business in Zambia had to cross the border and were often arrested in Zambia. They needed to pay to get out of jail and recently had to deal with new passport requirements that would cost them more money. Rather than needing to use their cell phones to stay in contact with WITC in the urban area 75 kilometers (46.60 miles) away, these women needed to use their cell phones to coordinate their business operations with other women in their rural home area. It was only by working together that these women were able to continue businesses in their home area while also needing to travel to Zambia, sometimes for a week or more, to bring back goods they could buy there for less money than in the DRC. As I listened to these stories, I was humbled to be in the company of such strong women and I told them so. The Congolese businesswomen told me how much it meant to them that I had come to visit them and hear their stories. We shared a moment where our hearts came close together. I hoped that my visit would somehow bear good fruit. All these years later and so many miles apart, I often think of these women and the dialogue we shared that warm day. Even if we couldn't completely bridge our divide, I felt that not all divides are bridged by physical structures or even formal communication.

The second story from my visit takes place on the day before I left Katanga, when I went with my adoptive family group to visit Fungurume Mining Company, a trip of about 175 kilometers (108.74 miles) in a large SUV over many kilometers of bad roads at high speeds. We left at 6:00 in the morning for the three-hour trip to Fungurume. After leaving the relatively good roads of the city, we traveled most of the way on red dirt roads through lush countryside punctuated with villages, one mid-sized town, and a military checkpoint. When we arrived at Fungurume Mining Company (FMC), we were greeted at a company gate by professionally friendly security guards who rode with us to the office where we were to meet our host. After touring the mining operation—one of the richest copper deposits on earth—we were joined by representatives from an international development NGO and drove to visit two small businesses in Fungurume that manufacture goods to sell to FMC. At lunch we talked with representatives from the international development NGO and FMC about their programs working with local people to develop sustainable businesses and build financial assets. We shared information, good food, and mango pie, then got on the road back to the city early enough so we would be at our lodgings before dark. At the time I didn't realize how significant this meeting would be for the text-messaging project.

The next day I left my friends and colleagues in the DRC and flew home via Johannesburg. After 37 hours of travel, I arrived in the subfreezing Minnesota winter and blowing snow just in time to start the information design class for spring 2010. I contacted people I had met from the international development NGO and asked if they would like to work with my class and our original partner on the text-messaging project. I was delighted when they said they would, especially when I subsequently learned that our original partner did not feel that WITC had the capacity to continue working on this project with us. Suddenly we had a new partner for the text-messaging system and needed to refine our project focus for this new situation. With our new partner, we determined that the text-messaging system should be targeted not to the businesswomen in Kasumbalesa, but to small farmers and artisanal miners near Fungurume, primarily to deliver pricing information for maize, cobalt, and copper.

As simple as this text-messaging system sounded, my students and I quickly realized that there were numerous complications to designing such a system. My students had many questions and although I answered them as best as I could, based on nearly three years of work on the project and a recent trip to the DRC, I knew my understanding of the situation was incomplete. We regularly emailed reports and questions to our international development partner in Fungurume and incorporated his responses into our design process. After each iteration of this process, we realized that we had more questions and so repeated this cycle of question, response, and design adaptation throughout the ten weeks we worked on the project. Although we began the semester working with university engineers and primarily considering technical issues involved in the design—how many bits can be included in an SMS message, Swahili syntax, standardized message format—midway through our work we hit the one question that would stop us for the rest of the term: whom would the small farmers and artisanal miners trust to provide local pricing information? One student in the class, who worked in the banking industry, applied a standard economic analysis to our project—whether potential buyers and/or sellers would be truthful in the information system or whether there would be irresistible incentives to cheat.

My students and I thought we were designing a technical communication system to deliver factual pricing information to small farmers and artisanal miners using their cell phones. We subsequently realized that we were designing a communication system that—before any of the technical aspects were designed—needed to be based on trusted relationships among the people using the system. In collaboration with our international development partners, the class recommended that we pilot test a very simple system that initially used pricing information available from African commodity exchanges instead of more local information. If we implemented this system to see what happened with a pilot group of trusted users, we could then continue getting feedback and redesigning the system with that initial group of users. We also hoped to develop relationships with university faculty in the DRC to develop a local team of information design specialists to take the lead on future iterations of the project.

## Hitting the Wall, Then Bouncing Back

Given the vagaries of life, politics, and commerce in Katanga, the mining company was no longer able to support its work with the international development NGO; they reassigned our international development partners to other locations. After three years of work, our project ended with no results and no prospects for continuing the project. I found myself questioning the possibilities of working with NGOs and companies in unstable circumstances. I thought of the strong businesswomen I had met in Kasumbalesa and wished I could work with them more directly. Questions of human relations, communication, and mutual understanding at the core of participatory design are thorny issues to engage. They require that those of us who wish to further our careers with other people's local knowledge be willing to give away something of ourselves in the transaction.

In the aftermath of this failed project, I tried to find some piece of the research that I could salvage. I returned to my basic knowledge of technical communication and looked for lessons about culture and communication that I could reclaim from the wreckage. I was not hopeful, but had to believe that there was valuable knowledge to internalize and share. I developed and presented papers at a number of conferences on topics ranging from rhetoric to the sociology of science. I wrote articles on this project, some of which were accepted and some of which were flatly rejected. I found some colleagues who encouraged me by expressing their interest in the project and lending their ideas for further exploration. I found other colleagues who rejected my authority to speak or write about Africa. I was not an anthropologist, an Africanist, or a sociologist. I was not African or African American. I was saddled with Victorian baggage of colonial "do-gooderism." I was accused of being insensitive and/or clueless. I was weighed down by my white skin, female gender, and older age. I was discouraged and almost silenced.

Then I received a particularly disagreeable rejection of an article submission to a journal and used my feeling of offense to steel my academic spine. What I salvaged from this failed project was this: My voice is authentic and my story is genuine.

In 2012 I joined the faculty at New Jersey Institute of Technology and was quickly asked to talk about my DRC project at a TEDxNJIT talk. Honestly, I was hoping to put this project in my rearview mirror but didn't feel that I could decline this invitation at my new job. I knew this project sounded exotic to most people and that I could add a new continent to the research offerings represented in my new school's public event. Once again I felt uncomfortable in being asked to represent the DRC and pose as an expert on anything there. Who was I to speak for Congolese people? Fortunately, I had some start-up funding and decided to use it to invite a graduate student from my Congolese host family to speak with me at the TEDxNJIT event. Yvan Yenda Ilunga accepted this invitation and we presented our talk titled "Trust," which focused on the supply chain for minerals mined by artisanal miners in the DRC—many of whom are children—through middlemen buyers and large mineral commodity companies, to cell phone manufacturers/sellers and the phones in our hands. Our talk discussed how to establish the trust needed to ensure fair mineral transactions in the context of corporate social responsibility. It also highlighted the idea that each of us who buys a cell phone has a responsibility to the child miner who will never hope to own such a device. I'm not sure that's the message that the administrators at my new job expected to hear; it wasn't an epideictic of praise for technology like that which characterized the other talks that evening. But it was the message that came from my experience, my heart, and from my collaboration with Yvan and his family.

After our collaboration in 2012, Yvan returned to South Africa and earned his master's degree. He came back to Newark for doctoral studies, and we started to plan a new phase of the price-sharing project focusing on farming cooperatives rather than the more fraught mining sector. And then the DRC politics intervened to make this project impossible. Ongoing separatist sentiment in Katanga flared into street fighting even in Lubumbashi. The country's president sought to neutralize Katanga's political impact by breaking the mineral-rich province into four new provinces with four new governors. Families and other cultural groups were divided, but conflict continued. In Kasumbalesa, where we might find farmers and businesswomen to work with, cross-border transit was often threatened by conflict between factions in Zambia and the DRC. The political climate in what was Katanga Province was not safe enough for either Yvan or me to consider working there with community partners. No depth of knowledge about technical communication could overcome this political obstacle. I think about those strong women I met in Kasumbalesa and wonder how they are doing.

## Suggested Readings

Buchanan, R. (2001). Human dignity and human rights: Thoughts on the principles of human-centered design. *Design Issues*, *17*(3), 35–39.

Buchanan discusses the concept of "human-centered design," emphasizing that we should not lose sight of the "human" focus in this term, even as pressures to increase efficiency and instrumental usability impinge on design processes—especially in technological designs.

Longo, B. (2014). R U there? Cell phones, participatory design, and intercultural dialogue. *IEEE Transactions on Professional Communication*, *57*(3), 204–215.

This article also covers the project described in "Accidental Tourist," raising questions about participatory design for information and communication technologies (ICT) projects when collaborators do not "speak the same language:" How can communication researchers effectively build trusted relationships with colleagues in developing nations in order to facilitate successful participatory design projects?

Longo, B. (2014). Using social media for collective knowledge-making: Technical communication between the global north and south. *Technical Communication Quarterly*, *23*, 22–34.

Drawing on a case study growing out of a colloquium on technology diffusion and communication between the Global North and South, the author proposes that technical communicators be attentive to the participatory nature of social media while not assuming that social media replace the dynamics of face-to-face human interaction.

Longo, B., & Ilunga, Y. Y. (2012). *Trust* [Video]. TEDxNJIT. https://www.youtube.com/watch?v=Q3_nqibpkjU

This video of the project described in "Accidental Tourist" includes perspectives on the DRC project from the author and Congolese collaborator Yvan Yenda Ilunga on the ethics of artisanal mining, supply chain transactions, and mobile phones.

Rönkkö, K., Hellman, M., Kilander, B., & Dittrich, Y. (2004). Personas is not applicable: Local remedies interpreted in a wider context. *Proceedings of the Participatory Design Conference 2004*, 112–120.

Processes for technological design that do not include the perspective of potential users tend to result in devices and products that reflect assumptions of the people who have the power to decide what to do to whom (Moses & Katz, 2006; Rönkkö et al., 2004). It is only by including voices of people whose identities, lives, and social positions will be affected that designers can hope to maintain a focus on the human beings in human-centered design. And in doing so, designers can affect social and cultural changes that potentially benefit people throughout the social strata.

## Discussion Questions

1. How would you have responded to the student who suggested giving iPhones to the businesswomen, small farmers, or artisanal miners in Katanga Province?
2. What issues does this case raise concerning research design in transnational technical communication?
3. What roles does trust have in conducting transnational participatory design projects involving technical communication? How can researchers and community partners build trust?
4. What roles do economic and political conditions play in conducting transnational participatory design projects involving technical communication? Why should researchers incorporate these effects into their projects?
5. What roles do local informants play in transnational participatory design projects involving technical communication? How do researchers and local informants build trust as a foundation for these projects?

## Author Biography

**Dr. Bernadette Longo** is an associate professor in the Department of Humanities and Social Sciences at New Jersey Institute of Technology. She is the author of *Spurious Coin: A History of Science, Management, and Tech-*

*nical Writing* (State University of New York Press, 2000), *Edmund Berkeley and the Social Responsibility of Computer Professionals* (2015), and *Words and Power: Computers, Language, and U.S. Cold War Values* (2021). She is also the coeditor of *Critical Power Tools: Technical Communication and Cultural Studies* (State University of New York Press, 2006) and coauthor of *The IEEE Guide to Writing in the Engineering and Technical Fields* (2017). Dr. Longo has also written and presented numerous journal articles and conference papers. She currently enjoys life by a small lake in New Jersey.

Chapter 5

# Across the Divide

## *Communicating with Company Stakeholders in Papua New Guinea*

Bea Amaya

Although I could not understand most of the words being spoken, I began to get the idea that there was a problem. I was walking around the periphery of the crowd—this collection of shareholders who had come out to hear the company presentations—and was taking pictures and greeting villagers. But something in the tone and timbre of this speaker, addressing the crowd in a language that I could not understand, caused warning bells to start going off in my mind. I began thinking that perhaps I should casually make my way back to one of the "caged" vehicles we had arrived in. Some of the villagers appeared to be challenging the speaker's remarks, and voices on both sides were increasing in volume. As I brought my camera up to take one more photo of the speaker, I noticed the leather jacket he wore on this hot day opened just enough to reveal a holstered weapon. While this image was still framed in my camera lens, his hand reached down to grab the weapon. This is about to get bad, I thought. Really bad.

## The Road to Papua New Guinea

In the fall of 2009, I brought one of my daughters into my database development consulting business to help take over my existing projects. I was several years into a multiyear contract with "Big Oil Co."[1] but was becoming concerned that the need for the services we provided were becoming more maintenance than development focused, an indication that it might be declining. So I started taking on some different kinds of projects. One project came to me through one of my husband's contacts during the time he was working his rotation in Escravos, Nigeria. The contact asked if I would be interested in helping a small tribally owned company, PNG Landco, located in Papua New Guinea, build a business plan for providing supporting trucking and transportation services to the area's growing oil and gas sector. Of course I was interested. Actually, I was more than interested, I was thrilled. By the end of the year, I was asked to join the team in a management position that required my in-country presence.

Papua New Guinea (PNG) is a small country occupying half of an island—the other half containing the Indonesian provinces of Papua and West Papua—just north of Australia. The country is noted as being one of the most culturally diverse in the world, with more than 1,000 distinct ethnic groups and over 800 languages spoken in a nation of only 8 million people. It is also a highly rural nation, with 87% of Papua New Guineans living in rural areas, and there is even some evidence that there are still a few tribes that have never been contacted by outsiders. But for me, PNG is the country that forever changed my perspective on what it means to be a global citizen.

## Living in PNG

I spent a great deal of time researching the country and its people before making the move and was tremendously grateful for those researchers, missionaries, anthropologists, teachers, and so many others who had taken the time to write about their own experiences and share them with people like me. But, of course, living there was the ultimate learning experience.

1. The name of the actual company has been changed.

My business partners were all PNG nationals, so my expat experience in-country was quite different from the other expats I came to know there. For example, the largest communities of expats were from the near-neighbor countries of Australia and New Zealand. I found that those groups typically lived in "expat housing," which spanned in range from the small and cramped worksite housing to luxury apartments with swimming pools, views, and beautiful furnishings, but always including stringent security protections. For my accommodations, I chose—against the advice of expat friends in country—to live in a PNG neighborhood. In spite of the dangers and lack of security protection of this choice, it was among the best decisions I made there, allowing me to experience life from a very personal and local perspective and to provide my neighbors with access to me and my own cultural perspectives as well.

For my first year there, I lived in one of six two-bedroom "condos" in a community outside the capital of Port Moresby. I had no hot water available for the first 30 days I was there . . . *Brrr!* There were people of all ages in the complex—at least half of the small apartments containing large extended families—with varying degrees of English-speaking skills. But we made do. My most regular and frequent visitors were a pair of 10-year-olds—a boy who spoke no English and a girl who was quite fluent. We played games, worked puzzles, worked on my pidgin language skills, and filled pages in the coloring books I had brought with me. Whenever I was home, they were my constant companions.

One of the items I brought into the country with me was a projector that I could hook up to my television in order to watch movies projected on the wall. Weekly gatherings at my place on Friday nights became the norm for the complex, when everyone from boisterous young neighbors to shy older security guards would gather at my place to watch rugby matches on the wall and cheer their teams on. It was great fun, and a great time for cultural crossings as I struggled to understand the rules of the game of rugby and they learned to be comfortable in my presence.

My "schedule" at this time was very different from the typical expats who went home every week or two for a break. My schedule allowed me to go home once a year for 30 days. I did, however, take frequent weekend trips to Australia, as this was a relatively short hop away. While it was certainly still a foreign country for me, I did have access to things there that I had been used to in the US, like fast food, hair salons, movie theaters, trains, buses, and boats, and as long as I did not speak in my south Texas drawl, the anonymity that was never available to me in PNG.

It was on these excursions out of the country, and in a more pronounced way on my first trip home, that I discovered I was changing. I found myself in situations outside of PNG that, for the first time in my life, made me feel uncomfortable and out-of-place. For example, because of the overwhelming curiosity about me in PNG, I had developed the local habit of speaking to people I did not know, looking people in the eye and smiling or nodding my greeting, and noting body language that indicated whether the people around me were "from my tribe" (more about this later). I developed two little speeches that I played in my head when crossing from one space to the other. When traveling out of PNG I would recite the following:

> You are leaving your cultural comfort zone so remember:
>
> > Do not look people in the eye for more than a few seconds.
> >
> > In an elevator or crowd remember to distance yourself physically.
> >
> > Do not speak to strangers unless spoken to.
> >
> > Do not play with, or touch, children without permission.
> >
> > Do not be alarmed that you understand the conversations—all of them—around you.
> >
> > And remember, nobody sees you.

There was, of course, a reverse list that played in my mind when returning "home" to PNG that went like this:

> You are returning to your new cultural environment, your home, so remember:
>
> > Look people in the eye and acknowledge them so you will not look "suspicious."
> >
> > Do not move away when people stand close to you in an elevator or crowd.

> Speak a few words to anyone who appears to be interested in you.
>
> Do not be afraid to play with, or touch, children you encounter and expect them to want to touch you, especially your skin and hair.
>
> Listen for English being spoken around you and acknowledge the speaker with a look or nod as people are sometimes doing it for your benefit or simply to "show off."
>
> And remember, everyone watches you, everyone sees you, and you are always on display.

For my second year, I moved into an apartment on the top floor (11th) of a business building in downtown Port Moresby that was only a few blocks from my office. This allowed me to walk to work easily. The apartment came with a large collection of empty coffee cans available to be placed strategically for water collection when it rained. While the apartment certainly had problems, it did afford me some amazingly awesome views. I was also across the street from a hotel with a separate restaurant and bar that could get very rowdy in the evenings, so I learned a bit about PNG street fighting and drunken brawls as well. It was during this period that I also learned about my mistaken assumption about "PNG culture." Despite the relatively small population in the country, there is really no such thing as PNG culture. It is always "cultures"—rich, complex, diverse, and intricate multiple cultures.

When I began walking to work every day, I took at least four trips each day. I walked from my apartment to the office in the morning, from my office to lunch at noon, back from lunch to the office, then from the office home in the evenings. And, of course, there were walks to the shops, office supply store, and even to the surrounding hotels where get-togethers with coworkers and others were frequent. I was only in residence a few days in the capital before I realized that many of the people I encountered on my walks knew exactly who I was. "*Monin' Boss Meri*" (Good morning, madam) was the greeting I received most often. It was always the "boss" part that confused me most, but seemed to be a form of showing respect for me. The facts that I was *tall*, *older*, *white*, and a *woman walking* the streets of the city made me known to all. The other fact that it took me a while to understand was that people in PNG were used to identifying others by their cultural or familial connections. After rumors that I was one of the many wives of the company's managing director (my boss) died down, the idea that I was connected by family ties died as well. Questions asked on the street indicated a great deal of interest about why I was in country and what my connection to PNG was.

In the middle of my first year in PNG I was given a gift of a lovely *bilum* from the Highlands Region. A *bilum* is a woven bag carried by many Papua New Guineans for the purpose of holding items and leaving the hands free. Men often wear short *bilums* around their necks and hanging on their chests, to hold small items such as money or *buai* (betel nut mixed with lime and chewed and spit much like chewing tobacco). Women generally wear larger *bilums* across their bodies, or, when carrying heavy items including babies in them, across their foreheads and hanging down their backs. I have many memories of seeing even older women walking along the roads in bare feet and, with a *bilum* strap across their foreheads, carrying unbelievably large loads of firewood on their backs. So, the *bilum* is an all-purpose and indispensable utility item and mine was no different. It was large, colorful, and I wore it strapped across my body and hanging at my side. It carried my billfold, phone, a notepad and pen, and any number of random items, and I wore it everywhere I went, both in country and out.

Because the woven pattern of my *bilum* was recognizable as coming from the Southern Highlands province, I became known on the streets of Mosbi as the *hailens meri* (Highland's woman), a name that indicated my "tribal" affiliation. The impression that I got from friends in the city was that Southern Highlanders were often considered by others as difficult, belligerent, and quick to anger. So, they were not respected in the city so much as feared. Although I obviously did not fit this characterization, with my living, shopping, working, and interacting with the "locals" on a daily basis, my association with the Highlanders seemed to afford me a bit of protection. And although I could not pick out from the people around me just by looking which were accepting of me, I could certainly tell by their actions which accepted, which feared, and which disliked me for this affiliation. "Belonging" in this way, this obvious and yet internal way, is something I still struggle to explain to others and something I have never experienced to such an extent outside of PNG.

## Trucking and Transport

Soon after moving to PNG, I took on the management position at PNG Landco that involved tapping into my technical writing skills and I found myself writing everything from high-level policies and procedures to field-level standard operating procedures, forms, flowcharts, checklists, and more. Although my knowledge of the trucking and transport sectors was lacking, my experience in the oil and gas industry was considerable and put me in a unique position to add value to the team.

The country of PNG is unique in many ways, but one of these is that the rugged terrain and endless rain-forests work together to inhibit travel. So, these same natural geographical features that are responsible for the rural lifestyle, the isolation of tribes and villages, and the resultant language diversity, are also features that contribute to the greatest challenges in the movement of goods across the country. Our truck drivers were primarily engaged in moving goods from the port city of Lae to the oil and gas fields of the mountainous regions of the Southern Highlands, as well as transporting materials across the worksites within those regions. This work could be highly dangerous, involved long hours on the road, and often involved risks that were difficult for me to even comprehend. The only activity that I participated in that gave me an inkling of what our drivers deal with on a daily basis came with my road trip for my organization's annual general meeting.

## The Annual General Meeting

During my first year in PNG I had been tasked with creating the company's first ever Annual General Report. It was a modest document that was largely made up of basic financial reports. In the second year, we decided to attempt something a little more complex, and more in line with the reports of the companies we were trying to impress, namely, Big Oil Co. and PNG Petroleum. My first step when crafting our document was to review their reports to get a good feel for what the expectations for the document were. My analysis indicated the document should introduce the major leaders in the company and discuss their roles, provide both a report of the most recent year's financial progress and the short-term vision for the future, and include a brief reference to the company's place in the larger picture in PNG. I was also tasked with the unusual chore (at least in my mind) of creating the first share certificates, and an associated database to record the member data, that the company had ever issued. Again, my procedure here was to review other company share certificates and ensure ours were in keeping with these "norms."

The drafting of the Annual General Report document was rather straightforward and involved meetings with the company chairman, managing director, and other members of the Board of Directors. This group of 11 Papua New Guinean men included two from other provinces and was made up of several with university degrees, several with considerable business experience, and several who were primarily important village leaders. And although I was raised as a small-town Southern Baptist with no interest in drinking, I learned to *pretend* to enjoy the occasional local brew and smoke a cigarette now and then as one way to connect with these men, especially the ones who spoke little or no English, to earn their respect, and to assert my right and intention to *belong* in the company, in the management team, and in the country.

Despite the university-educated among their ranks, the company leadership was still strongly influenced by its primarily oral culture. I believed this idea of creating a document with photos and quotes to be novel to some of these leaders and honestly did not expect it to be of much value to them or to the shareholders. But by the time we had the first draft printed and stapled, the managing director and I realized the document itself had the potential to be more valuable to us than we first anticipated. So, after a few weeks of meetings, planning, and discussion, the leadership team decided to create 1,000 copies of the completed document and take a "road trip" to deliver news of the health of the company at a dozen or so meeting sites with "shareholders" across the Southern Highlands Province. I had traveled to the region on several occasions, visiting our worksites in Kutubu, Mendi, and Tari, but I could see this trip was going to be something special.

I spent the next few weeks finalizing the document details, collecting and organizing the financial information, and negotiating with the local printer to create a professional document. During this entire process, my focus was on making the board members proud by presenting a document to Big Oil Co. and PNG Petroleum that would impress them so much that they would be eager to enter into further business discussions, and ultimately, agreements with us. In short, my target audience was really two people—our contracts administration contacts at these two companies. It was during the first meeting on the first day of our journey that I began to realize I had made a mistake in identifying my *primary audience*.

There is no feasible way to travel from the capital of Port Moresby to the worksite in Kutubu, our first meeting location, other than by air, so our team members traveled to Kutubu early in the morning with the first meeting planned for midday at a PNG Petroleum campsite known as the "Summit." The shareholder structure at PNG Landco was not at all what I was used to in that it involved several tiers of individuals who pooled their money to invest in the company. The bottom-tier shareholders were typically villagers who had contributed a large portion of their very small incomes to the venture. Additional tiers in the structure included families, clans, and tribes in complex arrangements that I was never able to fully understand. The first group of shareholders we met with at the meeting on the "Summit" was from the surrounding area and consisted of mostly middle-aged men who worked at or near the campsite. We gathered in an open-air pavilion where documents were distributed, and presentations made—by the managing director, general Trucking Operations manager, and general Finance manager. The presentations were made in the PNG version of pidgin known as *tok pisin* (talk pidgin) and were well-received by the shareholders. During the presentations, I performed my normal duties of photographing the proceedings and interacting with (smiling, greeting, shaking hands, maintaining eye contact, etc.) the members in attendance.

It was during this very first presentation that I noticed something unexpected. Over the years I have received several general annual reports and attended a few annual meetings. These previous experiences never prepared me for the attention that was paid to the actual, tangible, physical documents we had created. Upon receiving the documents, the shareholders examined every page, studied each of the faces of the Board of Directors and the Management team, looked at every photograph, and even ran their palms over each page of the document, seeming to take pleasure in the *feel* of it. I never intended for something I considered to be a straightforward business document to be treated with such, well, reverence. I was confused.

Once the presentations were concluded, share certificates distributed, and time spent engaging in answering questions and visiting was finished, we moved our group down to the base camp at the bottom of the mountain in order to prepare for the journey to the next site. After a few hours of preparation, the vehicles were all assembled, and our caravan was ready to move. The vehicles were all rugged escort vehicles with colorful markings showing the company name and logo, flashing lights, heavy-duty suspensions, and finally, the items that I can only describe as "cages." The cages were heavy wire mesh grids that were screwed in place across every piece of glass on the trucks and were designed to protect the driver and passengers from the road attacks which, in PNG, often involved attempts to break through windows using large rocks as projectiles.

Our second meeting was held around an arrangement of old wooden tables at a kind of low-rent roadside travel lodge for local travelers and again involved a small group of men with little-to-no English-speaking skills. And once again, I noticed the odd response to the physical documents we distributed. As men who came late to the meeting noticed the documents, they were quick to ask for a copy of their own. I even noticed several men taking an extra copy of the document to carry home. The time spent examining the documents was much longer than I anticipated, involved more close examination than I had ever seen over a business document, and involved much pointing, smiling, and touching of the document itself. But it was not until the third meeting that I began to think of a possible explanation for the unexpected behavior.

For the third meeting, we drove at an unusually high speed on the rough roads through the jungle to arrive on the shores of Lake Kutubu right at nightfall. With flashlights and headlights to help us, we climbed into dugout canoes that were very large and deep—they carried eight people with all their luggage easily—to travel across the lake to our destination on the peninsula, arriving in extreme darkness then climbing along the slippery path from the shore to the top of the hill. By the time I found my place in a typical PNG bed (i.e., not soft) in the corner of a small room and covered myself to avoid the unbelievably large palm-sized moths attracted to my light, I was sore, drained, and had not had more than a few words of understandable conversation for a whole day.

The next morning found me up at sunrise and when I stepped out of my cabin and made my way to the meeting place, the view I was greeted with left me speechless. Our vantage point looked over the island and the lake and was peaceful and breathtakingly beautiful. This is probably the most awe-inspiring view—a large expanse of lake and jungle where no hint of habitation could be seen in any direction—that I have ever encountered. It was amazing.

When the meeting time neared, I could not believe the number of people—30 or so—who had made their way to what seemed to me an unusually remote location. Yet, here they were, and here was that same odd behavior when interacting with my documents. As the meeting commenced and the presentations began, I began to think about the unexpected phenomenon I was observing. Why were these people, so many of whom could not read a word of the English text from the document, so interested in them? Why did they hold them so prominently and visibly while I took photos? Why did they handle them so carefully? And why did they take such care of the documents, and carry them back to their village homes, after the meetings were concluded?

I was finally forming a hypothesis about the documents that I continued to test along our journey. And while I can never be sure of the accuracy of my hypothesis, it was enough to adjust the way I think about *primary* and *secondary audience* when crafting documents and an example I continue to use in my university classrooms when discussing the topic with students.

Over time, I have come to believe that the documents became, for these people, a kind of tangible evidence of each person's worth or value, and of their place, their right to belong, in the organization. Even though these people were not organizational members and were not workers that received a regular paycheck or regular communication from company officials, the document they held in their hands tied them to the company in a way that I had trouble comprehending. As more meetings were held and more observations made by me, I continued to be amazed by what I was seeing. And when the meetings became more crowded, and there were fewer documents to go around, I could see obvious envy in the eyes of those who did not receive their own copies, and pride of ownership by those who did.

## Culture and Conflict

A common experience during my time in PNG involved incidents in which the culture that I was trying to respect collided with my own cultural experiences. One example was the time I chatted with two women discussing the fact that the husband of the woman who was a bank manager did not want her to wear pants as, in his opinion, it made her appear "manly." As the conversation progressed, I learned that the husband did not have a job or any income. He insisted on driving his very successful wife to work every morning and picking her up at the end of the day. Although she was perfectly capable of driving a car, he feared that her doing so would reflect negatively on him. I got the impression that the woman was disappointed by this, but also that she agreed with his assessment of the situation. Listening to the story and engaging in the conversation without expressing my own opinion, my expressly cultural view of the situation, meant that I was able to learn so much without offending or insulting the others. It was a technique that was so obviously beneficial that I began to work at honing the skill.

Another time I experienced an internal culture clash came during a quarterly management team meeting when the topic brought up was about one of our workers at the Kutubu base camp who had been released by the doctor to come back to work after an injury. The problem? He and his coworkers all believed he had been "cursed," and local management was worried about the impacts on the workforce if he came back to work. So, after a lengthy discussion, the company decided to retire the man and to pay him a regular salary to stay at home. A year before, I might have scoffed at the idea or been appalled at such a "business decision" by educated men. At this meeting, however, I was so surprised to find myself in full agreement with the team. And it was not because I placed any value in the idea that the man was actually cursed, but rather that I recognized that no matter how I feel about it, the fact that the workforce believed it was a strong motivation to take this unusual business action.

The incident with the gun that happened on the annual general meeting journey was another such cultural event. We had arrived at the meeting site where one of the largest groups of villagers we had met with had assembled, around 100 villagers, in an area known for violence even among Southern Highlanders. Many of the people in attendance had traveled great distances, walking for hours to get there on time. As I made my way around the edge of the crowd, I was surprised to see how many machetes were being openly carried at the meeting.

The meeting occurred in an open area between some temporary buildings that were up off the ground with areas beneath them that were open. The local leader and our business leaders stood on the porch of one of these units while the crowd congregated in front. There were many women and children in attendance at this meeting, making it different than most of the others. The women and younger children gathered under the structures to enjoy the shade as the day was quite hot. I was walking the periphery taking photos and greeting people as the meeting began. It did strike me as odd that the local leader was wearing a heavy leather Australian-styled duster on such a hot day. But I had seen other odd clothing choices before, especially among the local men, so I saw no reason for concern.

The meeting started with some opening remarks by this local leader and when it appeared that he was coming to the end of his speech, someone at the front of the crowd addressed him directly in a way that seemed challenging. Although I could not understand what was being said, I had the distinct impression it was about the distribution of shareholder funds among the villagers. When several additional raised voices joined in, I glanced to see how far away I was from the caged vehicles and remember thinking it might be a good time for me to head that way and take a break.

I raised my camera with a telephoto lens up so I could take one last photo, when I noticed the leader's jacket fall open, got a look at the holstered gun, and saw the angry look on his face as he reached to grab it. I felt positive that no matter what the outcome, this was going to be a once-in-a-lifetime event for me, this witnessing of a weapon being drawn at a company shareholder meeting.

Before the gun could clear the holster, however, I saw the company managing director, my boss, who was standing next to him, quite gently place his hand on the man's arm and speak quietly to him. At that moment, everything changed. The angry look was gone, the tension in his arm was released, his hand moved away from the weapon, his jacket fell back over it, and the angry voices were silenced. In a few minutes, the meeting proceeded as the others had with presentations, questions and answers, distribution of shareholder certificates, and general smiles, handshakes, and photographs. Even I found myself shaking hands with, and smiling at, the gun-wielding leader before packing up to go. It was not until we were back on the road that I noticed a collective sigh of relief from those in my vehicle. It seems everyone believed it to be a very close call.

## Lessons for a Lifetime

My time spent in Papua New Guinea was one of the most life-changing experiences I have had the pleasure of enjoying. I learned to pay attention to those around me and to listen, truly listen, without judgment to what they wanted to tell me. I learned to react with pleasure when a stranger wanted to "practice their English" on me, or when a child reached up in awe to touch my face or hair, and even learned to laugh at young boys who would run up and touch me because they had been dared to. I learned how to kindly and quietly decline when someone tried to treat me differently, like attempting to let me go to the front of the line, merely because of my white skin and their long practice of deference to it. I also learned that I was forever changed and would never again experience that blessed and blind sense of thinking I always knew better than those around me because of my lucky upbringing in the US.

My time in Papua New Guinea also helped my professional development. I learned that, as a company manager, what workers perceive is as important as what I believe. I learned to hone my technical communication craft when discovering how difficult it could be to code-switch in the appropriate places when writing Australian-English documents. But most of all, I learned to think about the documents I create by focusing not only on who I intend them for, the primary audience, but to pay more particular attention to who else might engage with them or be impacted by them, the secondary audience.

There were many "great divides" between me and the people I briefly encountered or the people I dealt with every day in PNG. I was white; they were not. I was a female in a leadership position in the company in a place where this was rare. I spoke only one language well, English, while most of the people I encountered every day spoke multiple languages but had little or no grasp of English. And I had an education, had traveled, and lived in a nice and comfortable home, while many of my PNG friends had none of these experiences. And yet it was I who felt so much richer every time I found a way to reach across to the other side and touch, and be touched by, another's cultural experiences. It was I who felt changed by the touch.

## Suggested Readings

Carbaugh, D. (Ed.). (1990). *Cultural communication and intercultural contact*. Lawrence Erlbaum Associates.

For a nice collection of cultural, intercultural, and cross-cultural encounters and their associated "communication patterns," the Donal Carbaugh's book is a great, if somewhat dated, resource.

EMTV Online. (2017, May 23). *Opening of new classroom in remote part of Nipa Kutubu* [Video]. YouTube. https://youtu.be/9LBVg6hLShE

This short video provides a brief but accurate depiction of the people; terrain; school buildings; English language use; pidgin (Tok Pisin) use; mixture of traditional and modern dress; and habits such as hand-holding among men, chewing of sugar cane, sitting on the grass during outdoor meetings, etc. of the area of Kutubu where some of my story takes place.

Hunsinger, R. P. (2006). Culture and cultural identity in intercultural technical communication. *Technical Communication Quarterly*, *15*(1), 31–48.

Hunsinger focuses on the identities that people bring to communication encounters and how important it is to understand these impacts. The difficulties in cross-cultural communication are examined here and new ways of approaching cultural issues in technical communication are suggested.

Smith, L. T. (1999). *Decolonizing methodologies: Research and Indigenous peoples.* Zed Books.

Smith's book is a "cautionary tale" told from the colonized Indigenous perspective and provides a counterstory to Western ideas about research and the pursuit of knowledge. It can be an uncomfortable read for Western researchers, but the value of exploring this perspective cannot be underrated.

Wardlow, H. (2006). *Wayward women: Sexuality and agency in New Guinea society*. University of California Press.

Wardlow's book is a detailed ethnography focused on Huli women, "passenger women," who accept money for sex. It is a rich and detailed examination of culture, relationships, desires, engagement with modernity, escape, and more.

## Discussion Questions

1. What kind of preparation would you engage in if you were to be moving into a very different cultural landscape for study or work?

What kind of people would you seek out for a trustworthy assessment of this landscape?

2. In the discussion of the bank manager's story, do you find it plausible that she could accept wearing dresses and being driven to work by her husband every day without objection? Do you think this is an aspect of cultural conditioning, or something else entirely? What questions might you have asked her had you been there?

3. Do you find the author's discussion of her reminder lists when travelling in and out of the country to be reasonable? Do you think this procedure helps or inhibits the transitions? Why?

4. The author admitted in her story about her *bilum* to becoming attached to the item and goes so far as to talk about how she carried it everywhere she went, even when out of the country. Why do you think she carried it even to places where its value—being recognized, offering protection, offering a feeling of belonging—would not be realized?

5. In the story about the handgun at the shareholder meeting, the author appears to be surprised by her own reaction (or perhaps lack of reaction) to the incident. What do you think it is that surprises her? What do you believe your reaction would have been in the same situation?

## Author Biography

**Dr. Belinda (Bea) Amaya** is a systems analyst at a company that manufactures HVAC equipment for industrial and commercial customers in Houston, Texas. Although her primary duties involve data collection, analysis, and reporting, she also continues to write occasional technical documents as well. Her academic interests are primarily in three areas: project management—she holds a Project Management Professional certification; intercultural communications—especially in the workplace; and data visualization—her current favorite is Microsoft's Power BI. She also teaches university-level undergraduate courses, currently at the University of Wyoming, and believes the space occupied by adjunct instructors like her to be valuable to students as well as to the instructors themselves.

Chapter 6

# "Nuesta vida en el medio oeste, USA"

## *Listening to Mexican Immigrants*

Laura Pigozzi

Entering any immigrant community as a researcher can be challenging. Even a carefully considered research design does not protect one from missteps. A research participant in a study I once conducted explained how to create an environment where members of the Latino[1] community would be comfortable talking with a researcher (me). He advised the following:

> So yeah, I mean, yeah, I think people will talk but it, because, [it] depends on the attitude of the interviewer is how you can [*can't hear*] especially ahh if its someone that, that ahh who looks like us you know? [*researcher: affirmative sound*] that is not dressed so fancy [*researcher: affirmative sound*] that looks like me [*researcher: "that you can sit down and talk to*"] I see someone with rings and all that stuff and I feel intimidated and I know others will feel intimidated. Well not me, I won't . . . —Research Participant

1. I have chosen to use the term *Latino* and *Latina* instead of *Latinx*. Those are the terms, along with *Hispanic*, used by members of the community where this research was conducted.

With respect to his criteria: I do look Latina (because I am) and I was wearing unremarkable clothing. However, I was also wearing diamond wedding rings, two bracelets with diamonds, and a heavy gold-and-silver bracelet. I wore this jewelry regularly outside of the research setting; however, I was thoughtless to wear this jewelry to the interviews—loudly proclaiming and affirming socioeconomic differences. He was being polite by not referring directly to me in his comments and by saying, "Well not me, I won't . . ." he was reassuring me that though I had committed this faux pas, he would continue to speak with me.

I always believed, somewhat unconsciously, that my ethnicity and life experiences naturally informed how I approached community research and allowed me to relate to and understand research participants. My mother is first-generation Mexican-American. Her parents, my maternal grandparents, immigrated to the United States to escape the Mexican Revolution in 1915. My father was third-generation Irish-American. I was raised in close interaction with my maternal Mexican family, and this ethnicity has informed my choice to focus on the immigrant Latinx community in my academic research. My hope was, and continues to be, that my work will help that community, whether directly or by increasing insight into the lives of these immigrants.

How and what to research was a question I pondered as I progressed through my PhD coursework. My training in bioethics (my doctoral minor) had informed me that minorities are underrepresented in medical clinical research. Having Latinos participate in clinical research is essential since limited participation results in limited data. Inclusion of all races and ethnicities increases the generalizability of research results, which is critical for the elimination of health disparities.

The specific research project I will be describing, which involved three studies over the course of three years, looked at Latino immigrants' comprehension of informed consent information. Participation in a clinical trial requires individuals to take part in an informed consent conference. In this conference the potential participant is given a detailed explanation of the study and, among other things, told the potential risks and benefits.

Comprehension can be difficult for recently immigrated persons or foreign-born individuals who have limited English language skills and/or limited formal education. Other social determinants may compromise individual autonomy, a bioethical tenet. Lack of comprehension, or incomplete or faulty comprehension, threatens the validity of consent. If researchers are committed to comprehension of trial information (as the Belmont report charges), they must make sure prospective participants are truly informed and understand all elements of the trial. Though quite a bit of research has been done on consent, I could find none that focused on Latino immigrants as clinical trial research participants.

In this research project, I made a distinction between what I termed as "legal consent" and "moral consent." By legal consent I was referring to consent garnered by a clinical researcher following their Institutional Review Board (IRB) requirements. These requirements often do more to protect the institution than the participant. Moral consent, in contrast, means the clinical researcher has ensured that participants comprehend all information fully. This requires a knowledge of the participants' cultural values and beliefs. It also requires an investment of time.

In the end I believe I not only answered my research questions, but provided a nuanced description of one community that can help other researchers. I succeeded, in part, by drawing on the cultural understanding I had by virtue of my own Latino heritage. Moreover, I also succeeded because I chose a research method that pushed me to reflect on my place within the research, and because I employed the principles of participatory research.

## Planning the Research Design

When I conducted the research described in this chapter, I was attending a large, public, Midwestern research university. I began by looking within the medical school for an ongoing clinical trial that I could join with my project. This proved more difficult than I anticipated. I sent out many emails over the course of many months and most were ignored. The literature affirmed that it was hard to "piggy-back" onto existing clinical trials as it added unwanted complexity for the clinical trial researchers. Moreover, I was looking for Latino trial participants, but there were few due to the geographical location.

During the time I was contacting clinical trial researchers, I attended an academic conference where I was introduced to a methodology called *analogue participants*. The presenters, who created this concept, explained that it could be used when one did not have access to actual clinical trial participants. Analogue participants watch a video of a person presenting a consent conference. This seemed like a viable solution. As I looked further into this methodology, I noted the major criticism of this methodology was that it might create unrealistic scenarios. For example, asking someone to imagine they had a stage-four cancer and were being recruited for a clinical trial was not going to produce genuine results. I considered this for a while and ultimately identified a way to address the criticism: a well-patient study. Most people could imagine themselves being recruited into a well-patient study that examined, for example, elements of a healthy lifestyle.

I located an ongoing well-patient study sponsored by my university's school of public health. The public health researchers were much more receptive to my inquiry than the medical researchers. They generously shared all the study materials (which were already translated into Spanish). Though not labeled as such, the study was basically a childhood obesity study directed at the Latino community. One of the enrollers from the trial agreed to film his consent conference for me to use. I was almost ready to begin Study 1. Now all I needed were participants.

## Researching in Sacred Spaces

I had solved the problem of how to locate Latino immigrant participants in a previous study I had done for my master's degree thesis. That study looked at the effectiveness of existing healthcare materials created for the Latino community. The idea of approaching Catholic parishes as recruiting sites came about when I was discussing that project with community healthcare organizations that work in the Latino community.

My decision to approach Catholic parishes that served predominately Latino populations[2] as my research sites was carefully considered. Faith is essential in the lives of these participants. For many their faith community is not separate from their Latino community—it is central to it. I recognized an obligation to demonstrate respect and honor the values of this community while in their sacred places.

I initially contacted priests at two different parishes via email asking to meet. At Parish 1, the priest and I had in common a mutual friend, which aided in establishing rapport. A priest at Parish 2 already had, coincidently, an interest in the research topic. He wrote:

> Hola Laura,
>
> As a former mental health practitioner, I am extremely interested in this issue. I have wondered about that since my start here as

2. The majority of these parishes and the participants in the studies referenced in this chapter were of Mexican heritage.

> a pastor five years ago. How is informed consent understood, if at all, by immigrant Latinos, and what are the ethics of garnering a signature when the consequences are not understood?
>
> You are welcome to conduct interviews at [Parish 2]! I'd be happy to get together with you to discuss this in more detail. I may invite others of our staff to be present if possible, too.
>
> Thank you for this overdue research!
>
> (name withheld to maintain anonymity, personal communication)

I was transparent with the community leaders (priests and other parish staff members) about the research questions I would be exploring and the details of the research design. These conversations included discussions about why this research was important to me and how the results might be used. At both parishes, I had many email exchanges, as well as face-to-face meetings, to discuss recruitment details, how best to introduce the study, and how and where to conduct the interviews and focus groups. Parishes have full calendars, so identifying available dates was challenging.

Additionally, I knew that many in this community were unauthorized immigrants. Conducting research through the Church, I believed, provided some degree of protection for unauthorized participants, as it was an accessible and familiar location. I felt a great responsibility to protect the identities of the participants. At the time this study took place, there had been two recent Immigration and Customs Enforcement (ICE) raids at local daycare centers. One participant explained how this had affected the community: "Estamos muy estresado al respecto. Por todo lo que ha sucedido. Los grupos vigilantes, la inmigración, la reforma, el tratamiento policial, grupos vigilantes . . . todo eso afacta a la comunidad Hispana de una manera u otra."[3] To protect the anonymity of the participants and the parishes, I do not use any names, locations, or other identifying markers in this chapter.

In the spirit of community research, I followed all suggestions I received. For example, one parish suggested I hold focus groups after Mass, instead of enrolling people for individual interviews. The community

3. "We're very stressed about it. Because everything that's happened. The watchful groups, the immigration, the reform, police treatment, watchful groups . . . all of that affects the Hispanic community in one way or another." Translated by the author.

leaders felt people would be more comfortable speaking in groups. This was confirmed by a research participant, who noted, "Y cuando el grupo esté unido, tendrán más confianza así que cuando estén con más Latinos tendrán más confianza."[4]

## Recruiting

Recruitment details differed slightly for each parish. At Parish 1, I posted flyers in the church the week before I was introduced to the parish. The community leaders reviewed the flyer before posting and suggested changes. On the Sunday the project was introduced, I attended and participated in the Mass. I am a practicing Catholic and I wanted the parishioners to know this, hoping to make them more comfortable. At the end of the Mass, the priest encouraged parishioners to consider participating in the study. I spoke briefly from the altar and then the interpreter I worked with, Emilce, explained how to sign up for an interview time.[5] We had set up a table with forms, pens, and candy for the children in the foyer, and as people approached us, we explained to potential participants that we wanted their opinions about a medical study. Interested community members left their names and phone numbers so we could call to remind them of the date and time of their interview. To maintain participant's anonymity, the sign-up sheets were destroyed immediately after the interviews took place. We conducted interviews over two weekends in the church rectory.

At Parish 2, I also posted flyers, which were approved by those community leaders, the week before the research took place. At the end of Mass, parishioners were invited to participate in a discussion (focus group) and share a meal in the church basement. Those who were interested simply came downstairs. Parish 2 was also where the third study in the project took place. Recruitment for that study took place in the same manner.

Having community leaders introduce me and the research study raises power issues. Prior to conducting the research for this project, I had discussed using a Catholic parish as a research site during a PhD methods seminar. The professor asked me if I was exploiting the community. At the time I could only answer, "I would never do that!" Even now I don't

4. "And when the group is together, they'll have more confidence so when they are with more Latinos then they'll have more confidence" (translated by the author).

5. I speak Spanish but do not consider myself completely fluent. For that reason, I worked with an interpreter—a native Spanish speaker.

really have a better answer. I can only say I identify as a Latina, and I am a practicing Catholic. I have relatives who belong to the same demographic groups as these community members. Consequently, I did all I could to treat the research participants with dignity and respect.

## Working with the IRB

To begin any research project involving human participants, the Institutional Review Board must grant approval. I did not have difficulty obtaining IRB approval for this project and to further protect participants' identities, I sought and was granted a waiver for signatures on my research consent forms.

However, issues with protocol changes began early in the research. For Study 1, I had originally stated I was going to interview 20 people. That number had been arrived at by reviewing published studies, but I did not adequately consider the applicability to my study. As explained, in this first study I recruited participants by having them sign up for an interview to take place the following week. I suspected that not everybody would attend their scheduled interview. What I did not anticipate was that the interviewees would arrive with other family members (sisters, mothers, aunts, cousins), as well as with friends and neighbors. The interviews evolved into small focus groups. On the very first day of interviews, I reached the 20-person limit, even though I had planned two weeks at that location before moving to the second location.

I spoke with the IRB, explaining my situation. They recommended that I submit a Change of Protocol form and request 200 participants. They explained that it did not matter if I did not interview 200 people; they cared if I interviewed more than approved number. It was at this point that I made an appointment to speak with someone in the IRB office who reviewed applications. I explained to the reviewer that I was using grounded theory methodology and explained how grounded theory worked. I also advised the reviewer that I would be submitting more Change of Protocols requests during the course of the project and asked that he be the one to review them so I would not have to continuously explain why I was making changes. This worked well and expedited the time the Change of Protocol requests took for approval.

IRBs are quite concerned that giving gifts to participants may be coercive and are very cautious when approving such gifts. One unexpected protocol change, among several, that needed approval is explained in the following excerpt from the Change of Protocol request:

> The second change that is being requested is to offer $15 cash versus a $15 [local grocery store chain] gift card. When purchasing some the gift cards used in Phase 2 [research at Parish 2] the bank thought the transaction was fraudulent and deactivated the cards without contacting me. Some of these deactivated cards were distributed after one of the focus group sessions. When I learned this I brought new cards for those participants, but trust was broken. Since Phase 3 will be conducted with the same community I feel it best not to continue to use gift cards but rather offer cash. The monetary amount remains the same.

This situation necessitated not only a protocol change, but work to restore trust. That work took the form of conversations with the participants who received the deactivated cards. I apologized, explained what happened, and gave them cash. Word got out that I had repaired the problem and trust was, for the most part, restored.

## Conducting Interviews and Focus Groups

As described, the locations were chosen because they provided a convenient and familiar site for the participants. Hospitality was offered in the form of food and the presence of a childcare provider. As a token of appreciation for their time, the participants were given a $15 gift certificate to a local grocery chain (which evolved to $15 cash, as explained in the previous section). All communication was done in Spanish.

We—the interpreter and I—greeted the participants as they arrived and offered them refreshments. We assisted in getting the children settled. To maintain the participants' dignity in consideration of possible low literacy, we read the consent materials for this study out loud, periodically asking for questions.

The script we (I and the interpreter) used was localized and anticipated areas that might be confusing. I tried to address these confusions preemptively. One probable confusion point was consent for this study—the participants were consenting to be in a study about consent. We did our best to explain this. Another point that was important to be as clear as possible on involved the use of the analogue patient methodology. The participants needed to understand that they were not being enrolled in the public health study that was the stimulus. They were *pretending* that José (the enroller in the video) was enrolling them, but it was not real. We used the word *imagina* (imagine) quite a bit. The localization of the script was important. One participant noted that "how" a person is asked is significant, observing, "Usted, usted debería estar muy impresionada con cuánta gente, cuánta gente en la comunidad quisiera hablar con alguien, y quizás no slo sobre una cosa pegueña, pero sobre muchas cosas y nadie numca, numca se molesta en preguntaries, ni en averiguar en cómo preguntaries."[6]

For the first two studies, the research stimulus materials were already in Spanish (these were materials from the actual public health study), though I believe they could have been better localized. The research stimulus materials for the third study (a diabetes clinical trial) were in English. I had these translated into Spanish. The reason for the change in stimulus is beyond the scope of this chapter.

Interviews and focus groups were audio-recorded, not video-recorded. Though I did take photos of the rooms where the focus groups and interviews took place to document the room arrangement, I published no photos of the exteriors of the buildings, and no faces of any participants.

During the focus groups and interviews we made no reference to legal status. However, at times we were asked directly by the participants if unauthorized people could participate. We assured participants that legal status was unimportant to us. However, at times reactions to the stimulus reflected uneasiness surrounding an unauthorized status. One participant remarked, "Pero suena muy racista y aparte ya lo aclaro ahí [referencing the research stimulus], ya dijo que ninguna agencia de gobierno federal tendrá acceso a esta información. Entonces, se escucha muy racista para mí. Yo porque notengo documentos no puedo participar en un studio? Se escucha muy mal."[7]

Emilce, my interpreter, and I were familiar with the expectations and communication norms of this culture. However, to avoid stereotyping I

6. "You, you should be very surprised how many people, how many people are out there that want to speak to someone and maybe not just for one little thing but on so many things but no one ever, ever bothers to ask them that and no one even bothers in knowing how to ask them" (translated by the author).

7. "But it sounds very racist and he already clarified here [referencing the research stimulus], he said that any federal agencies will have access to this information. So, it sounds very racist to me. Only because I don't have legal documents I can't participate in this study [the study being discussed in the research stimulus]? It sounds very bad" (translated by the author).

fore-fronted the medical notion of *cultural humility* in order to focus on the individual. I once heard a Native American activist talk about working with others to provide medical care for those in his community. He used the term *cultural agility* to refer to the practice of being open and flexible and aware of the complexities of culture.

I followed best practices for creating interview and focus group questions. For the most part, things went smoothly. The participants seemed at ease and had interesting discussions. However, in hindsight, I think piloting the scripts would have served to better localize the questions. Furthermore, when reviewing the transcripts, I can see that I did not do an adequate job of explaining why I was conducting this research and how it could ultimately benefit the Latino community.

## Addressing Power Imbalances

There were certain strategies and actions that I deliberately included in the project to redress power imbalances. The literature suggests that racial concordance between researchers and participants provides a level of comfort and trust. I, along with my interpreter, provided racial concordance with the participants. While I am proficient in Spanish, it is not my first language and I do not consider myself fluent, especially with colloquial language. Therefore, an interpreter, who was a native speaker, accompanied me for all interviews and focus groups. Emilce served as an interpreter, translator, and most importantly, became a friend. As a child of migrant farmworkers, she provided many insights into the community, increasing our collective cultural competency. She could relate to the community members' many struggles and pointed these out to me as further explanation for their responses. The participants responded to, and I'm sure appreciated, her compassionate demeanor. This research also provided language concordance in that all communication, written and spoken, was in Spanish.

Another action that works to level power imbalances and gain trust is to demonstrate your familiarity with the participants' culture. Creating a culturally appropriate environment not only allows participants to feel comfortable, but it also demonstrates respect. The locations for the interviews and focus groups provided a safe, familiar environment. At both sites I arranged for a Spanish-speaking childcare provider to be present and included child-friendly snacks and toys. I also served food. For meals, I cooked the food myself. I believe this helped dissipate the power imbalance and conveyed appreciation for the participants' time and opinions. If a meal was provided for a focus group session, the recruitment flyers clearly stated that information.

These choices stemmed from my cultural knowledge. I knew that people attended church with their families, so it was more than likely that their children would accompany them to the interview or focus group. In order for the participants to concentrate on the research conversation, the children needed to be occupied in a way the parents were comfortable with. My research design reflected the knowledge that family and community are held dear. One participant remarked, "Mas que nada, nuestros hijos son el bienestar de nuestra familia."[8]

I did not spend a significant amount of time researching what might be most culturally appropriate in this research design, relying on my innate knowledge. I did, however, consult with the priests, parish administrators, and staff who worked directly in the community. They all agreed the design was a suitable approach.

## Reflection

I recently read the book *American Dirt* by Jeanine Cummins, and then read the subsequent backlash, which caused me to reflect on the ethics of my research. The book tells the story of a mother and son's journey from Mexico to the United States, as they escape drug cartel violence. It was initially heralded by some as the "great immigrant novel," until the book was roundly criticized by Latino authors and activists. The book, the critics noted, does not accurately portray the immigrant experience and worse, is full of clichés and negative stereotypes.

Not only was the book denounced, but Cummins herself was criticized for speaking for Mexican immigrants. Her maternal grandmother was Puerto Rican, and according to her critics she has, in the past, identified as white. She is not a migrant, and she is not Mexican. David Bowles explains the significance of this in a Medium blog post titled "Cummins' non-Mexican Crap," "Why does her identity even matter? Because she gets nearly everything *wrong* as a result." He asserts that the book is exploiting brown trauma for "the white gaze and white book clubs."

8. "More than anything, our children are the well-being of our family" (translated by the author).

Among the many problems Cummins's critics have with her ethos, is her privilege. The literature exploring "speaking for others" also places researcher privilege in a problematic role. The literature warns that privileged persons speaking for those with less privilege, or for oppressed groups, may reinforce stereotypes or oppression. If academics speak for others, they must speak with careful consideration for the *effects* of their work.

This controversy caused me to seriously reconsider my research. I sincerely hope my interpretations accurately portrayed the experiences of the community members who participated in my studies. Interestingly, though I did not intentionally examine the issue of "speaking for others" while planning and executing the research, grounded theory allowed me to recognize many potential pitfalls. Reflexivity was one such tool. Moreover, I had done a fair amount of reading on participatory research. Though the project I'm writing about was not formal participatory research, I kept participatory research principles in the forefront. For example, I periodically checked my data interpretations with community leaders.

"Parachute research" refers to research that leaves nothing behind for the community. It is the opposite of participatory research. A researcher enters a community, gathers data, and then leaves. I wanted to avoid this as much as possible. After the studies were complete, I met with the community leaders to talk about the overall research results. I also discussed, in detail, the specific categories that had emerged from the data, along with their implications. In some cases, programs were initiated, or were already in place, to address some of the obstacles or struggles identified—for example, providing English language classes or health screenings by public health organizations. We discussed what the parish might do in the future in direct response to the research results. These conversations were how I attempted to leave something behind for the communities. Reflecting on the *effects* of this research, I believe I did what I could to genuinely bring forward the voices of those communities.

## Suggested Readings

Alcoff, L. (1991). The problem of speaking for others. *Cultural Critique*, *20*, 5–32.

Alcoff, a philosopher, writes thoughtfully on the question, "Is the discursive practice of speaking for others ever a valid practice . . . ?" (p. 7).

The importance of the context of a discourse is one such issue as certain contexts are associated with structures of oppression. Possible responses to the problem are discussed and Alcoff then provides four sets of interrogatory practices to help evaluate speaking for others. To conclude, Alcoff urges readers to question the effect speaking for others may have: "will it enable the empowerment of oppressed peoples?" (p. 29). Though now nearly 30 years old, this article remains relevant on the topic.

Beauchamp, T. L. (2011). Informed consent: Its history, meaning, and present challenges. *Cambridge Quarterly of Healthcare Ethics, 20*, 515–523. http://doi.org/10.1017/S09631801110000259

This article introduces the reader to informed consent. Beauchamp, a philosopher well versed in bioethics, presents the history of informed consent and discusses its present meanings. The most common meanings are grounded in institutional and regulatory roles. He ends with a discussion of the contemporary challenges.

Charmaz, K. (2006). *Constructing grounded theory: A practical guide through qualitative analysis.* Sage Publications.

This textbook should be read by those interested in using constructivist ground theory in their social science research. It is accessible and provides illustrative examples throughout.

Israel, B. A. (2001). Community-based participatory research: Policy recommendation for promoting a partnership approach in health research. *Education for Health, 14*(2), 182–197.

This article discusses community-based participatory research (CBPR) use in public health, listing the key principles and rationales. It also includes policy recommendations intended to advance the use of CBPR and illustrates some of the challenges encountered when using CBPR.

Noe-Bustamante, L., Flores, A. (2019). *Facts on Latinos in the U.S.* https://www.pewresearch.org/hispanic/fact-sheet/latinos-in-the-u-s-fact-sheet/

The Pew Research Center provides well researched and reliable facts on the 60 million Latinos in the United States, including those who are undocumented. It is a good place to start to understand this population. The site includes, among other topics, the population distribution, educational attainment, English proficiency, and poverty rates.

Ornelas, I. J., Yaminis, T. J., Ruiz, R. A. (2020). The health of undocumented Latino immigrants: What we know and future directions. *Annual Review of Public Health*, *41*, 289–308.

This article summarizes the literature concerning how an immigrant's experiences during and after migration affect mental and physical health. The article also provides recommendations for future research, which includes some best practices for working in unauthorized immigrant communities. The article ends with some recommendations for public health practice.

Pigozzi, L. (2018). Negotiating informed consent: *Bueno aconsjar, major remdiar* (it is good to give advice, but it is better to solve the problem). In L. Meloncon & J. B. Scott (Eds.), *Methodologies for the rhetoric of health & medicine* (pp. 195–213). Routledge.

Those interested in reading more about the research discussed in this chapter can consult the chapter listed above. In it I first discuss informed consent's place in the field of bioethics, as well as describing how I view consent through the lens of rhetorical theory. I then discuss the research project's methodology and end by detailing some of the unique results of this project.

Pigozzi, L. (2020). *Caring for and understanding Latinx patients in health care settings.* Jessica Kinsley Publishers.

This book is intended for healthcare providers who wish to know more about how Latinx immigrants *may* view health. I stress the "may" in order to recognize that not all immigrants hold these beliefs, or hold them to varying degrees. The book discusses cultural and health beliefs, the role of religion and spirituality in health, mental health, communication strategies, and interpretation and translation.

## Discussion Questions

1. What is your feeling about "speaking for others?" Do you believe the researcher must be racially and experientially concordant with research participants? If the researcher is not, do specific steps need to be taken?

2. Have you ever examined your implicit biases? One way to test for them can be found here: https://implicit.harvard.edu/implicit/aboutus.html.
3. The process of informed consent is historically rooted in law and policy. How might this inform a difference between legal consent and moral consent (as defined in the chapter)?
4. Would you consider using community-based participatory research? What might this method help you uncover that is not currently explored in your research design?
5. How would you address the ethical implications of conducting research in a religious institution such as a Catholic parish?

## Author Biography

**Dr. Laura Pigozzi** is a visiting assistant professor in the Cook Family Writing Program at Northwestern University. She holds a PhD in rhetoric and scientific and technical communication with a doctoral minor in bioethics from the University of Minnesota. Additionally, Dr. Pigozzi has affiliate faculty status in the School of Nursing, University of Minnesota and affiliate faculty status in the Center for Community Health Equity, DePaul University/Rush University. Her research interests include the rhetoric of health and medicine, technical and professional communication, and cross-cultural communication. Dr. Pigozzi's work incorporates rhetorical, quantitative, and qualitative analyses to build explanatory, interdisciplinary theories. This research contributes to a practical goal of accessible, usable, and effective healthcare communication. To that end, she has recently published a book for healthcare providers titled *Caring for and Understanding Latinx Patients in Health Care Settings*.

Chapter 7

# Syrian Refugee Women's Voices

## *Research Grounded in Stories and Recipe Sharing*

Nabila Hijazi

وَالَّذِينَ تَبَوَّءُوا الدَّارَ وَالْإِيمَانَ مِن قَبْلِهِمْ يُحِبُّونَ مَنْ هَاجَرَ إِلَيْهِمْ وَلَا يَجِدُونَ فِي صُدُورِهِمْ حَاجَةً مِّمَّا أُوتُوا وَيُؤْثِرُونَ عَلَىٰ أَنفُسِهِمْ وَلَوْ كَانَ بِهِمْ خَصَاصَةٌ

—Quran 59:9

As for those who had settled in the city and embraced the faith before the arrival of the emigrants, they love whoever immigrates to them, never having a desire in their hearts for whatever of the gains is given to the emigrants. They give the emigrants preference over themselves even if they may be in need. And whoever is saved from the selfishness of their own souls, it is they who are truly successful.

—Mustafa Khattab

"You are one of them; you represent their home—their origin and roots—and their future." These were the words of the local masjid's[1] Imam[2] when

1. A masjid (مَسْجِد) is a mosque, a place of worship for Muslims.

2. Imam (إمام), Arabic imām ("leader" and "model"), one who leads Muslim worshipers in prayer. *Imam* refers to the head of the Muslim community.

he invited me in 2016 to help welcome and work with Syrian refugees settling in our community. My efforts were genuinely and purely volunteer work, especially since Islam highly recommends that community members help each other. The main sentiment that prevailed at that time is that of "المهاجرين والأنصار."[3] The community welcomed a great number of Syrian refugees, who represented "المهاجرين": those who left their homeland to seek refuge from the oppression and persecution of the Syrian regime and the brutal civil war that consumed the lives of many innocent Syrians.

As a first-generation Syrian American Muslim female academic who immigrated to the United States in 1989, and who has successfully navigated and balanced the multiple communities I am part of the—American, Syrian, Muslim, academic—I was seen by the mosque's leadership as an ideal candidate the Syrian community can easily consult—an ally, a confidant Syrian refugee women come to for advice, whether religious, cultural, familial, and, most importantly, educational. Ultimately, I found myself in an ideal place to conduct research on the Syrian refugee community and the struggles they face living in a new country as they construct the diasporic home, as I am part of their lives and most intimate spaces: homes, religious circles, "coffee respites," and community events. However, when designing a research study, feminist scholars like Linda Alcoff warn against "the pitfall of speaking for others." I found myself in a challenging ethical entanglement: rich data and important voices that should not be ignored and a volunteer role that is paired with research interests, potential publications, and academic advancement.

Having a lot in common with my research subjects—Syrian refugee women—in terms of culture, language, dialect, religion, and identity struggles, most importantly struggles around constructing a Syrian home in a Western land, I was in a unique situation. I had some of the most intimate conversations with them, not only during my visits to their houses, but also in every stage of the research process. The project I will be describing looks at these conversations, how I prepared for and expe-

3. The term for migration in Islam is *hijra* (هجرة). The active participle of the word is *muhajir* (مهاجر). The term هجرة *hijra* involves permanent relocation to a land where one would feel stranger at first, suggesting vulnerability, hardship, and the need for help and guidance from the locals or those who took the journey before them. الأنصار means helpers or supporters. The term has Islamic significance. الأنصار were the local inhabitants of Medina who, in Islamic tradition, took Prophet Muhammad and his followers (المهاجرين) into their homes when they emigrated from Mecca during the hijra.

rienced them, and the numerous efforts I did in organizing adult literacy classes and in working with local mosques to collect funds and arrange after-school programs, including tutoring and Quran and Arabic classes, for Syrian refugee children. The numerous conversations I had with them in their homes and online, on WhatsApp, included rich, vibrant details that inspired me to reflect on my positionality as a researcher and the ethics behind my community work being convoluted by research interests. Being intrinsically linked to my research subjects and thus inseparable from their context, I reflexively examined my position as a researcher and the social, cultural, and personal forces that shaped it and reinforced the importance of ethical responsibilities with regard to conducting research within local marginalized communities.

## Research Inception and Design

In 2016–2017, while I was working on my PhD comprehensive exams, I was concurrently working with and volunteering for local mosques to help Syrian refugee families make our community their home. Being a member of a local mosque's board of education and due to my academic credentials, I was invited by the Imam to create adult ESL literacy classes for Syrian refugees. However, as I immersed myself in working with the local Syrian refugee community, I started seeing the potential of conducting research about the issues and challenges they face. I realized the volunteer work I was doing was multilayered and multidirectional, informing not only my religious and communal service but also my scholarly and academic work. Syrian refugee women's experiences back in Syria and here in the United States comprise a rich tapestry and fertile soil I could not ignore or avoid. The stories of Syrian refugee women surviving a ferocious war, a long journey—even bloody for some—and foreseeable struggles, struggles of existence, resistance, and survival, made me more invested in carrying out my research and bringing these stories and voices to light. However, my official, formal Institutional Review Board (IRB) research started with the ESL literacy class the mosque sponsored and I cocreated for refugee women to learn the English language.

Data collection for my IRB-approved study occurred in different stages. Before I started planning for the literacy class, I was involved in organizing community events at the mosque, welcoming the new groups of refugees and helping them settle in, furnishing their houses, helping with groceries, and translating documents. Due to language, cultural, and religious familiarity, I became close to many of them; they began expressing to me their interest in enrolling their children in Arabic, Quran, and Islamic studies classes. Working with two Islamic community centers, the masjid, where the ESL literacy class I cocreated took place, and another one in College Park, Maryland, I organized small fund-raising events and contacted family members and friends to raise money to cover the cost of these classes and to arrange for transportation, since many refugees at that time did not have cars or their driver's licenses yet. Through my interaction with them, I felt the urgent need for both men and women to enroll in literacy classes in order for them to become independent and manage their own affairs. Some were already enrolled in literacy classes through government-sponsored agencies, but many expressed their lack of interest, either because of the location or not feeling comfortable around strangers. Ultimately, through my conversation with the Imam and the mosque's board of education members, we decided to host literacy classes for both refugee men and women, since we saw their sentimental attachment to the masjid and their frequent expression of gratitude to community members.

## Site Description/Background

To solicit interest in the ESL literacy programs, I contacted two of the Syrian refugee women I had been in conversation with almost on a daily basis. Each lived in different neighborhoods that were known to house refugees. I visited both of them in their houses in December 2016. Both invited the Syrian refugee women who lived in the same neighborhoods to their houses in order for me to meet with them, soliciting their and their husbands' interest in ESL literacy programs and afterschool Quran and Arabic classes for their children.

With both visits, several women came and showed excitement about the programs and the tight Syrian, Muslim community they were becoming part of. Both women were very hospitable, serving Turkish coffee, Syrian sweets, and fruits. The living rooms were buzzing with women sipping coffee and conversing in Syrian, expressing the joy of the new bonds they were forming in this country. While all of the women identified as Syrian, they came from different provinces. The mosque and the relationships it was forging for them became the safe hub of their new life, while the community of "المهاجرين والانصار" became the seeds of a new, promising life and the pipes for new blood that is free of any racial contamination or biases. They viewed the masjid and the relationships and opportunities it created for them as invaluable. Coming to a new country and culture and leaving loved ones and entire lives behind was extremely challenging for them. Therefore, they jumped at any opportunity the masjid provided, with literacy classes being one of them.

With both visits, the women were socializing and enjoying their time, sharing feelings of joy because of the strong community support. They expressed their frustrations about how constructing a Syrian home in the United States was different than what they were used to in Syria. One woman mentioned how it was not possible for her to live the same lifestyle she had in Syria. While she has to honor her domestic and maternal roles here, she has to learn how to be more independent: drive, buy groceries, take her children to doctors' appointments, attend school events, and more. Their wifely and maternal roles meant extra layers of responsibilities they did not anticipate. Ultimately, learning English was a precursor and a necessity to achieve that level of survival and independence.

As the excitement winded down, I started recording the Syrian refugee women's personal information, including names, ages, education levels—both in Arabic and English—and home addresses. I asked for the same type of information about their spouses and children. I recorded the information on separate Excel sheets on my laptop. With these women being stay-at-home mothers and housewives, we decided to hold the women's class once a week on Wednesdays, from 9:30 a.m. to 12:00 p.m., when their children were in school. We agreed to have the men's class on Sundays, from 2:00 p.m. to 4:30 p.m. I wanted to have classes meet twice a week, but to ensure the majority could attend, I settled for once-a-week.

Based on interest and commitment, I drafted a budget plan that would cover the two teachers' compensations (one teacher for the men and the other for the women), the translator/teaching assistant's compensation, the transportation cost (transporting women to and from the mosque to their houses), and childcare cost. One of the Syrian refugee women offered to provide childcare services at her house during the women's class. She lived in the same neighborhood. She declined to attend the literacy class because her husband was disabled and she saw an opportunity for making extra money.

After all preparations were complete and classes were ready to run, all of the Syrian refugee men withdrew because many were working as taxi drivers, and attending class meant losing the extra potential income they would make if they were on the road. Ten out of the 15 women decided to attend; it was depressing, considering the effort I put in. We decided to cancel the men's class and proceed with the women's class, with the two teachers and the translator/teaching assistant. During orientation, I assessed all of the female students' English proficiency levels and decided to have the two teachers be in class, since the difference in students' literacy levels was strikingly clear. I created a WhatsApp group with all of the students' names and would send announcements and reminders in Arabic, Syrian dialect, on Tuesdays, encouraging all of them to be on time. And, even though I did not teach any of the actual lessons, I was always present at the mosque to greet the students when they arrived and as they left after each class.

We named the ESL literacy program RESLA, which stands for Refugee English as a Second Language Adult program. We chose RESLA because it sounds like the word (رسالة) in Arabic. This word has several positive meanings and connotations: its primary meaning is a written or typed letter that is sent in an envelope by mail. It connects to the idea of literacy and communication. We felt it would resonate with refugees since they are eager to contact their loved ones overseas. Also, I envisioned RESLA as an open letter about the research I am conducting on the underrepresented group of Syrian refugee women that has cross-cultural information that cannot and should not be ignored or overlooked.

RESLA became the major cornerstone of my research. I decided to interview the Syrian refugee women who enrolled in the class and regularly came to the local mosque for various reasons: praying, receiving charities and spiritual and emotional support, accompanying their children for Quranic and Arabic classes, participating in festive Islamic events such as Eid[4] parties, and attending ESL literacy classes—to name a few. The class ran during the spring 2017 semester. However, I could not conduct any of the official interviews until I had my IRB approved; it took longer than expected, because I was asked by the IRB to translate all of the recruitment forms, consent forms, and interview questions to Arabic since my research subjects had limited English proficiency.

4. Eid (عيد) represents Muslim festivals, in particular Eid al-Fitr (the religious holiday celebrated by Muslims worldwide that marks the end of Ramadan, the Islamic holy month of fasting) and Eid al-Adha (the second of the two Islamic holidays; it corresponds with the height of the Hajj, the pilgrimage to Mecca).

While initially I relied on the interviews for data sources, what became evident was that each detail, even minor ones, counted toward research. For example, the WhatsApp messages I had with Syrian refugee women on a daily basis, or the ones I sent on Tuesdays to remind them about being prompt and ready for the taxi drivers to transport them to the mosque facility for class on Wednesdays, proved to have a wealth of valuable information, from which I could glean details for my research. For instance, as the class progressed, some students provided me with excuses to miss class for reasons like doctor's appointments or frustration of not being able to progress in learning the language as expected. Their hesitancy in coming or canceling last minute were signs for me to pick up and eventually assess my research focus.

## Change of Heart: Change of Direction

Initially, the purpose of my research was to examine the challenges ESL literacy programs for refugees struggle with and to offer suggestions for improved adult literacy classes. These programs, even though valuable to many refugees, impose tricky and complicated situations, requiring commitment and consistency. I intended to evaluate the effectiveness of RESLA in meeting the needs, goals, and expectations of Syrian refugee women and to locate the best practices to teach this group. What became apparent to me, regardless of their lack of commitments, motivation, and enthusiasm toward RESLA, was that this group's literacy practices varied depending on the social and cultural contexts individuals came from. The literacy practices they brought with them do not necessarily conform to the Western definition of literacy that is equated with reading and writing. And, while I have a connection with this community, being a Syrian woman by origin, speaking the same language and dialect, and being acquainted with Syrian culture, my literacy experiences and approaches are not necessarily the same as theirs. I am an immigrant woman who came to the United States voluntarily long before they came, and I have a more extensive educational experience, in terms of higher education. Nonetheless, I exercise many of the literacies and gendered practices they did back in Syria and still do in this country: being a full-time housewife and mother and honoring all of the domestic roles expected of a Syrian woman, including cooking, pickling, or even being hospitable when guests come over. However, that did not necessarily allow me as a researcher to navigate and dissect the multiple challenges these women experienced while learning English. In fact, I realized that my relationship could overshadow what they wanted to say and even muddy the water. I had to step back and evaluate my research methods to understand the cultural practices these women brought with them and honor while living in this country.

## Research Ethics: Building Relationships of Trust

My previous experiences growing up and identifying with most of the issues and values of these Syrian women allowed me to relate to them, especially since I have not abandoned what I learned in Syria. I became a familiar face. The bonding experience with these women allowed me to ground my research in trust and respect. Nada, a Syrian woman refugee in her mid-thirties, whose husband became paralyzed due to war injuries, continually expressed and reiterated the sense of relief she felt around me because she perceived me as one of them, as I understood the struggles and challenges they were experiencing. However, coming to this program with a feminist lens, the savior lens, to help these women learn how to read and write to improve their educational, social, and economic status, I ignored the richness of refugee women's literacies which add to the multiplicity and diversity of literacies of this country. Their literacy and gendered approaches allowed me to see beyond my biases—to acknowledge how this group can speak for themselves and exercise their agency in ways that suit them. Instead of speaking on their behalf, I saw the need for them to speak and honor their voices and choices. They are the ones who decide what works best for themselves and their families, and they are the ones who should dictate the types of literacy that legitimize and advance their own and their family's status. This sense of agency became even more evident in the type of personal information these women were willing to share with me. Therefore, instead of one-time formal interviews, my research bifurcated into different sources and venues.

## Orienting Framework

I wanted to minimize researcher bias since I personally knew all of these women, and I worked closely with them, assisting even with family matters. I structured each stage of my data-collection process to allow these learners as much autonomy and agency as possible to select and describe their unique experiences with literacy, the challenges they've faced, and the benefits they've gained. By conducting oral interviews, I elicited students' perceptions of their own knowledge, strengths, challenges, and literacy growth through stories and oral responses. This method is especially important in studies of immigrant and refugee literacy practices, since many refugees do not have the strong educational foundations on which literacy is built. The oral interviews allowed me to avoid the pitfall of speaking for others and to "speak with" them. As an immigrant and a successful college-educated Syrian woman, I was already in a position of power; therefore, it was crucial that these Syrian refugee women speak for themselves, using the language they deemed fit to represent the intense details of their lived experiences.

Ultimately, I decided to use storytelling as a major part of my methodology because preliterate learners tend to communicate best through storytelling. The storytelling tradition in the form of concrete narrative sequences opened a window to shared context and culture. The storytelling approach allowed researchers to appreciate their research subject participants' cultural and racial diversity, be active participants in the broader conversation, and engage in academic inquiry. Appreciating the efficiency and importance of storytelling, I came to realize how this mode of data collection could be empowering to my participants outside academia, since preliterate learners often prefer storytelling over a more analytical interview because storytelling interaction is personal, engaging, and immediate. I accounted for and allowed my participants to present themselves as whole people and agentive storytellers. I attended to the different ways in which my Syrian refugee participants chose to represent themselves and demonstrate their agency, drawing attention not only to what was said but also to what was unsaid. Instead of dictating their answers, I sat and listened to their stories, allowing for spontaneous answers. Our collaborative question-response built a story about their experiences—past and present. This format helped them talk about their encounters in a new, individualized way that worked well for them and this study. It allowed them to express their frustration not just with language learning but also with their experiences fleeing war zones.

Rather than asking highly analytical questions, I assisted them in telling their stories in ways that would be comfortable for them. With each interview, I started with the first questions on the list; however, in a few minutes, it became apparent that the process was too rigid. Therefore,

I found myself spontaneously asking follow-up questions that emerged from the context of the conversations. The interviews switched to friendly conversations, with intimate details that only close friends would share over a cup of coffee, enabling them to begin restoring and reconstructing their lives in a setting of literacy, and allowing me to understand, digest, and analyze their motivation and challenges with learning English and their attitudes and goals toward learning and literacy. The conversations turned into a window to these women's own lives, including the struggles they faced and utilized not only in their literacy journeys but also in their constructing and reconstructing of a Syrian home in the diaspora. I utilized an in-depth narrative approach, allowing women's narratives and voices to be the focus of the study while capturing the complexity and richness of each woman's story, and highlighting their agency in storytelling and in the gendered literacy choices and practices they made to sustain their families.

During the interviews, we had flashbacks about these women's experiences in Syria during the war and how their literacy practices sustained them as they were held hostages in their towns and deprived of any supplies or amenities going in or out of their towns. One described how salt was a source of sustenance for them: preserving their food and healing their bodily wounds, figuratively preserving whatever is left from their lives. As she was describing how they used every grain of salt and the type of food they had when they were under siege in Syria, we occasionally diverged into discussing the gendered literacy practices, such as food making and preserving, we were taught at a young age. We would recount the recipes and the proper steps in making food that would not go bad regardless of temperature—foods like *labneh*,[5] cheese, *makdous*,[6] jams, or pickles. Sharing such recipes would lighten the mood as we would find ourselves choking when discussing the danger they witnessed or the loss they endured. The interviews and the research drifted from being formal to friendly to become even therapeutic.

---

5. Labneh (لبنة) is sack yogurt; it is yogurt that is strained to remove most of its whey, resulting in a thicker consistency than unstrained yogurt, while preserving yogurt's distinctive sour taste. Labneh balls are made by making labneh extra thick and then drying it under the sun to make it firm and last longer.

6. Makdous (مكدوس) is a dish of oil-cured aubergines. They are tiny, tangy eggplants stuffed with walnuts, red pepper, garlic, olive oil, and salt. Sometimes chili powder is added. Makdous is usually prepared by Syrian households around fall to supply for winter, and is usually eaten during breakfast, supper, or as a snack.

With each home visit and interview, I was welcomed with open arms; I felt at ease taking my *hijab*[7] off and even accompanying the women to the kitchen as they were preparing Turkish coffee for our interviews. During our conversations, we would check on each other, asking how our families, especially children, were handling the new environment. I shared tips and resources for raising children according to Islamic and Syrian guidelines, and they shared their frustrations regarding the environment their children were surrounded by. For instance, one complained that while she appreciated the volunteer services offered by government agencies, she expressed her frustration about female volunteers not properly dressed when coming to her house to work with her children. She was not comfortable around female volunteers coming to the neighborhood and her house wearing shorts or miniskirts. While it was none of her business, she considered the volunteers' dress code to be inappropriate, and she thought agencies that work with refugees should be trained to teach these volunteers about different households that function according to specific religious or cultural etiquettes. These types of conversations and stories are major threads in working with refugee communities, especially those that place high value on gender and its relationship to religion and culture. They shed light on future research practices that researchers can adopt or even consider when preparing to conduct studies among such communities.

## Reflection: (Re)designing Interviews as Relational Processes

Being aware of the danger of the asymmetries of power that can undermine the relationship with research participants, I invoked trust, respect, and transparency with my participants. I did not achieve that only through sharing a common language, dialect, culture, and religion with them, but also through my constant interactions with them—listening to their past and ongoing struggles, offering advice, acting as a mediator in communicating with religious figures in local mosques, soliciting and sharing information about educational programs for their children, sponsoring fund-raising events to cover the cost of these programs, and sharing information about

7. The hijab is a religious veil worn by Muslim women in the presence of any male outside of their immediate family. It should cover the hair, head, and chest.

multiple opportunities and events to promote and advertise their products.

Coming from a culture that is grounded in storytelling as an everyday practice, my participants preferred storytelling over a more analytical interview. For instance, with each interview, I would start by reading the questions I prepared, which made the conversation rigid, preventing the fluidity of the information and the richness of the details to the point that even they would point that out. For example, Umm Wasim[8] said, "You do not need to read all the forms and questions, just summarize the main point of the interview and the research and then we can engage in the conversation." Their reactions encouraged me to pivot away from the analytical interview process using the questions I prepared and instead to engage in an organic, natural way of storytelling and story sharing—allowing for some of the most profound, intense, and deep details to emerge. I broke the researcher/participant line and became comfortable with sharing and exchanging information that would be applicable to our diasporic lives and struggles in constructing a Syrian, Muslim home—opening the door for personal, private, intimate, and deep conversations. For instance, instead of reading the question, "Describe any struggles with coming to class? In other words, was it difficult for you to attend due to your gender and family role? How did you negotiate these difficulties?" I would make this statement in Syrian dialect, "ليه كان صعب عليكِ تحضري الصف؟," which translates to "Why was it difficult for you to attend RESLA class?" Starting with this statement, we would then dive into their domestic role and cultural and familial views on women's education. By not asking direct questions, I created the possibility that my participants would consider and share relevant ideas rather than limit their responses only to narratives that center around language learning. I opened the door for my participants to discuss with me unanticipated details that contributed to their gendered approaches to literacy.

In listening to them, I took the backseat to hear, listen, digest, and contemplate all the copious elements that made up their experiences and

8. In Arab cultures, mainly Syrian cultures and more specifically rural and suburban cultures, married women with children are addressed by the name "Umm" plus the name of the eldest child, mainly the male child. "Umm" means the mother of. If the family does not have male children, they use the name of the eldest daughter. Umm Wasim is the mother of Wasim. In these cultures, it is an honor to address married women with children this way.

honored their voices. In that way, I moved away from possible misrepresentation of their female refugee voice or from placing it in a state of invisibility. I engaged in authentic thinking and careful listening, where women's experiences are recognized as relevant data and evidence. I, along with my participants, used the language that represents them and their unique voices.

All of these strategies allowed me to move beyond the use of asymmetrical power relations. These practices opened a window into the women's hearts as they shared their experiences using their unique voice and language. Using their native language, the Syrian dialect, allowed me to preserve the rich details that could be lost through translation, especially when there is a third party involved in data collection and translation. By moving away from dominant ways of doing research and the rigid format of formal interviews, and by sharing my own experiences, including the struggles and the successes I have had in raising my children to be confident Muslim Syrian Americans, I crossed the dichotomy of the researcher/participant relationship and crafted research methods grounded in trust, respect, care, and empathy. By closely interacting with my research participants and assisting them in several matters, I was able to create not only a rapport but also a harmonious, friendly relationship with them. The relationship I built with my research participants went beyond a professional, formal level. It was immersed in stories and recipe sharing.

## Suggested Readings

Alvarez, S. (2017). Latinx and Latin American community literacy practices en confianza. *Composition Studies, 45*(2), 219–221.

Steven Alvarez argues to ground research in "trust, respect, and sustainability en confianza . . . [which] translates literally as 'confidence,' but in practice confianza means reciprocating a relationship where individuals feel cared for. [It] involves exchanging mutual respect, critical reflection, caring and group participation. Confianza is dialogical trust, acceptance and confirmation between researchers and communities" (pp. 219–220).

Booth, W. C. (2004). *The Rhetoric of rhetoric: The quest for effective communication.* Blackwell.

Wayne Booth argues that "listening will be useless unless you let it change your rhetoric" (p. 51). He emphasizes that "all good rhetoric depends on the rhetor's listening to and thinking about the welfare of the audience" (p. 54). The practice of "listening-rhetoric" draws attention not only to what is said but also to what is unsaid.

Foss, K. A, & Foss S. K. (1994). Personal experience as evidence in feminist scholarship. *Western Journal of Communication*, *58*(1), 39–43.

Feminist theorists Karen Foss and Sonja Foss emphasize the importance of "authentic thinking" and careful listening, where women's experiences are recognized as relevant data and evidence. Personal experience counts as valid data because it accounts for "women's personal narratives about the events of their lives, their feelings about those events, and their interpretations of them. They reveal insights into the impact of the construction of gender on women's lives . . . and their perspectives on what is meaningful in their lives" (p. 39).

Freire, P. (1971). *Pedagogy of the oppressed.* Herder and Herder.

Paulo Freire's notion is that marginalized members of society are the source of authority on their experience.

Glenn, C. (1997). *Rhetoric retold: Regendering the tradition from antiquity through the Renaissance.* Southern Illinois University Press.

Cheryl Glenn has cautioned against "gendered silencing." Here, Glenn emphasizes the need for feminine representations of agency to be vocalized and for silence to be recognized as a choice: "Silence is not necessarily an essence; it can be a position—a choice" (p. 177).

Glenn, C., & Ratcliffe, K. (2011). *Silence and listening as rhetorical arts.* Southern Illinois University Press.

This book is a collection of essays that demonstrate the value and importance of silence and listening to the study and practice of rhetoric. Respecting the power of the spoken word while challenging the marginalized status of silence and listening, the authors argue for acknowledging these overlooked concepts and their intersections.

Meyers, S. V. (2014). *Del otro lado: Literacy and migration across the U.S.-Mexico Border.* Southern Illinois University Press.

Susan V. Meyers argues that an activist methodology requires the "ethical responsibility of adding to or giving back" to local communities (p. 14). Meyers theorizes that a "reflexive critical ethnography" incorporates reflexive inquiry into its methodology. Researchers employing this approach position themselves as being intrinsically linked to those being studied and thus inseparable from their context.

Nussbaum, M. (1997). *Cultivating humanity: A classical defense of reform in liberal education.* Harvard University Press.

Martha C. Nussbaum argues that contemporary curricular reform is already producing such "citizens of the world" in its advocacy of diverse forms of cross-cultural studies. Nussbaum establishes three core values of liberal education: critical self-examination, the ideal of the world citizen, and the development of the narrative imagination.

## Discussion Questions

1. Being aware of the danger of the asymmetries of power that can undermine the relationship with research participants, how can researchers invoke trust and transparency with their participants?
2. When, if ever, is it legitimate to speak for others, especially for the less privileged? Should all speaking for others be condemned or just some?
3. Does research participants' lack of language proficiency justify the researcher's ability to speak on their behalf?
4. Is being a community member and researcher of that community considered speaking for others?
5. How do research practices transcend the boundaries of culture, religion, and gender? Does sharing similar aspects of the participants' intersectional identity allow the researcher to smoothly integrate into the community they are researching and capture some of the hidden important data?
6. Does being well-versed in the participants' gendered practices that are part of their everyday experiences allow the researcher to engage in common, pertinent conversations and in-depth research and data?

7. Does sharing the same religion, culture, language, dialect, and even struggles with research subjects allow for more authentic listening and research?

## Author Biography

**Nabila Hijazi** is a postdoctoral fellow at Loyola University Maryland's Writing Department. She holds a PhD in English language and literature with a concentration in rhetoric and composition from the University of Maryland College Park. In her dissertation, "Syrian Refugee Women in the Diaspora: Sustaining Families through Literacies," which received Honorable Mention in the 2020 President's Dissertation Award by the Coalition of Feminist Scholars in the History of Rhetoric and Composition, she draws on interviews and community-based work with Syrian refugee women in the Washington, DC region, to examine the cultural, economic, and political dimensions of their Arabic literacy practices and English literacy learning in the United States. Some of her publications include a chapter in the *Feminist Circulations: Rhetorical Explorations across Space and Time* collection, where she traces Muslim women's rhetorical tradition by tracking rhetorics that circulate and recirculate in the Middle East to rethink how rhetoric and religion circulate to a different context, temporality, and geographical location and relate to Muslim, gendered identity. Also, in her essay, "Bodies in Conflict: Embodied Challenges and Complex Experiences" (in press) in the edited collection, *Our Body of Work*, she takes up the notion of embodiment to analyze Syrian refugee women's experiences that make the physical body a source of knowledge. Her research interests include Muslim and refugee women's rhetorics and literacy practices.

Chapter 8

# Relearning Your Knowledge

## *The Loud Silence*

Yvan Yenda Ilunga

As I landed for the first time in the country that would become my home for several years, my first contact with a border agent quickly awoke me to the reality that I was facing an entirely new culture, and that there was no alternative to learning the language to participate in society. It was around 11:45 a.m. when my plane took off from the Democratic Republic of Congo (DRC), destination Johannesburg, South Africa. While on the plane, the pilot addressed us in both French and English; this was still manageable since the entire plane was packed with French speakers, which meant less intimidation communicating with one another. Once I arrived at the airport in South Africa, I could rely on the many signs that said, Welcome to South Africa, Please proceed this way for control, and so on. Less interaction made my life easy, but those comfortable moments of course did not last. After a short walk through the hallways of the airport, I found myself in front of a gentleman who asked, in English, for my passport and the reason for my visit. I froze, spoke unintelligible words, and looked at him straight in the eyes patiently. I was not afraid to interact, but I was concerned that my verbiage did not make any sense whatsoever. He noticed my struggle and understood that I knew what I was doing but was unable to express it verbally. That is when my struggle to try to break the silence began.

## The Beginning of the Dream

I am sitting in an auditorium with more than 800 students, patiently waiting for an extraordinary professor to introduce a course on theories of international relations. It is 2005, at the University of Lubumbashi in the DRC. The auditorium is packed, the students are busy discussing the credentials of the professor, whose name is extremely well-known among the top scholars of international relations in the DRC, though I am not sure what to think of him yet. Along with a few friends, we are waiting to see if the praises said about the academic background of the professor would match with his teaching. A few minutes later, the professor entered and introduced himself by saying, "After many years of traveling the world and working in the cabinet of the former president Laurent Desiree Kabila, and supporting the work of the ministry of foreign affairs, I am back to raise the next generation of internationalists." This fascinating introduction and way of approaching his class catapulted my attention and I could no longer afford to be distracted. Not only was he a very articulate professor, but his passion for the field of international relations was palpable, which inspired me above all else.

As I am sitting, completely captivated by his many narratives and stories from his trips to the UN Headquarters in New York to Addis Ababa, and so forth, little did I know, I had begun to dream of an international career as well. During this very inspiring and provocative first lecture, the professor passed around two books written in English (I cannot remember their titles). He wanted us to have a look and understand that most of the scholarly materials in international relations were written and discussed in English, hence the necessity to learn the language. From there, it became evident to me that to advance my knowledge in the field, I had no choice but to learn English. However, it was still too early to embark on that journey; I was only a second-year university student who simply had to write assignments and pass my classes. No international trips yet. But I was still allowed to dream, and so I did. I took pleasure in immersing myself in my academic training for five years with intensive research and theoretical knowledge, all in French. During that time, I had very limited exposure to anglophone scholars, and the education system in French gave me another important perspective in the field that I needed the most. In 2007, a cousin who was traveling from the United States to the DRC and I met in Lubumbashi to share ideas of local economic development with an American lady who was working with an international humanitarian agency in the DRC. The meeting revealed that all of us had the same passion for ending poverty. The meeting also revealed that our educational backgrounds were very similar but were completed in two different languages and worlds. At the end of the conversation, my new American friend handed me a book on poverty alleviation written in English. I must mention that our conversation was entirely in English as well! Of course, apart from the friendly initial greeting done in French, as a way of creating a harmonious space, the rest was in English.

To get back to the book, I took pleasure in contemplating the words in it, trying to make sense of some sentences and paragraphs, and getting an overall idea of the content. This was not easy to grasp; nor was it easy to discuss and share my ideas in English during the meeting. This encounter, and many more that followed, made me realize that I needed to learn English to be able to explain what I already knew in French. Otherwise, I would be the mute/silent intellectual in the room. This experience taught me that active participation in the construct of global knowledge would have to be supported by a new set of skills that I needed to grasp. Therefore, the struggle for the new language and the hope for an international career, informed by my academic background, became the two germinators of my dream.

## Networking and Small Group Discussion

The first test of a long conversation in English had been passed, and my interest in learning was becoming more evident. I had discovered my limitations and I could not stand another day without improving my English. As many scholars doing ethnographic study would tell you, stepping into the environment and embracing the culture are some critical steps for understanding a community. As an aspiring scholar, I also had to find a way to embrace the anglophone culture and way of doing life while in the DRC. I decided to increase my networking activities. Every time I knew that there was a scholar or missionary in town who spoke English, I would do my best to find time to meet and talk with them. These meetings were often concluded with an exchange of emails to keep the conversation going. Beyond the friendships that could result from such contacts, I was using these opportunities to improve my writing skills and knowledge of the language. With time, they improved.

During that same period of intense networking, I decided to join a Bible study group called Discipleship. This intensive discipleship training, theological in nature, was run by an American missionary affiliated with the United Methodist Church in the Congo. Our cohort was comprised of only seven people. Hence, this small group became one of my most productive platforms for practice. The training lasted a year and during that period I grew far more comfortable with my listening skills. Why does this story matter? To me, this is of great importance to everyone who is preparing to do international research and engage in cross-cultural scholarship. The first lesson I learned was that one must be prepared and familiarized with the culture, language and, if possible, the environment of the community or space where the research would be performed. This preparation could be conducted in the form of simulations or networking, as I did. In fact, over the years I have seen many good international projects fail because scholars/researchers were not culturally and linguistically prepared.

We live in a time and space in which it is effortless to gain knowledge of various communities with just one click on Google. There is no longer any reason for not being prepared. Though I was still aware of my insufficiency and lack of proper command of the language, deep inside I knew that there was no other way than to train and become prepared. I must also say that the people who were most advanced in language in my small group did not let me down. They did not mind my accent or bad grammar. They focused instead on the content of my message, which helped build my confidence. Many international researchers and scholars fail to perform well due to low self-esteem as a result of language barriers or insufficiency. If the academic community could work harder to give people the chance to be who they are without being judged, I am confident that the exchange of knowledge would be greater and more easeful.

Getting back to my story, this networking opportunity helped to keep me focused on the much-needed skills required to be effective in the space of international research and collaboration. I remember that my initial feelings in the small groups involved wondering if I were qualified to express my thoughts and share my knowledge. This is unfortunately an ongoing struggle that many people find themselves in, but it should not be an obstacle for engaging with the world.

## Jumping into the World

Jumping into the world led to the realization that there are systemic languages and verbal languages, which are often the basic tools for transnational and intercultural collaboration. While in the coming sections I focus on verbal language and the loud silence, I thought it would be quite important for me to devote this section to systemic language, which is here demonstrated through administrative and procedural research protocol approval.

In fact, after three years of multiple encounters, networking, reading and practicing my listening skills, it was time for me to jump into the world of transnational research and collaboration. My first international experience on transnational research occurred in 2010 on a trip to South Africa. With my basic working knowledge of English, which was later improved in my host country, and enough understanding of cultural differences between the DRC and South Africa, the first serious challenge was to adapt to the administrative procedures related to the conduct of international research. In many academic jargons, this would be the Institutional Review Board (IRB) processes. I remember one day, in the early afternoon, I was discussing with my supervisor in South Africa how to get my research fieldwork started. In my mind the process was easy, and I would be going into the field with limited clearance from the university. To my big surprise, getting clearance from the school would take about four weeks. This was because I had to work with human subjects and conduct international field research. At the hearing of this information, I started to realize that administrative processes also had to be considered part of the complexities and challenges of research. In addition, between the submission of my research proposal and approval by the ethics committee, I cannot remember how many workshops, one-on-one meetings, and conversations I had with my supervisor and other faculty members in preparation for fieldwork.

This fieldwork had to happen in the DRC, in a community I was very familiar with—which, for me, was like returning home. But little did I know that going home was one of the main concerns that faculty reviewers had with my proposal: "How could Yvan maintain the standard of his research while conducting fieldwork at home?" After some time, I understood their concern was valid, and that indeed I had to prove I was able to objectively conduct my research back home, in the region and community where I had many memories and personal ties. However, going back home to conduct research was an opportunity to retrieve my voice through many casual conversations with community members and participants in the research. It was also an opportunity to come out from my loud silence. The excitement of going home versus the question of objectivity in research led me to question whether it is fair to completely disconnect researchers from their feelings and personal views of issues they have either researched or directly experienced. This is one of the thoughts that I would like to honestly discuss with my fellow social scientists in more detail. It seems to me that we might be burying critical knowledge at the expense of preserving so-called objectivity in research and production of knowledge. This internal feeling of disempowerment and forced silence is real. I honestly still do not know how anyone can ever remain 100% objective (as defined in the methodology) in the social sciences while dealing with issues such as poverty, inequality, race, and exploitation.

Anyway, in that fieldwork, I had to make sure that I was objective. While I was struggling to figure out what pure objectivity would look like, since I didn't want to appear to be and act like an insensitive robot in the field, I had a very enriching Skype call with a friend of mine who helped me to figure this out. Although our talk was on another subject, she had mentioned to me that "so often scholars think that people in remote and underprivileged communities are not educated—the reason for the use of the term *such uneducated people*. This perception undermines the fact that these so-called uneducated possess other forms of education which are not very much formal. We need to be humble enough to hear them and consider their knowledge." This thought helped me figure out what objectivity for me would look like. It would mean listening and considering local knowledge without the pretention of knowing all or leading the thoughts of others in the production of the stories that I would expect to see coming from them. This worked well and procured a very successful experience.

However, not everything went smoothly; I guess this is a constant reality for many in academia. In fact, while the question of objectivity was addressed, the methodology was finalized and my research proposal was approved in South Africa, I had to get another approval from the DRC before starting the fieldwork. I believed it was still a fair process that would help to maintain academic rigor. But the challenge came because some in my academic institution in South Africa assumed that the administrative process in the DRC would be similar to the one in Cape Town. This misconception unfortunately continues to characterize many administrators in developed countries, who think that conducting fieldwork in foreign countries is as easy as it looks on paper. But remember, I had lived in the DRC for many years and worked with different scholars, public and private institutions, and so on. I had conducted fieldwork, developed community-based projects, established nongovernmental organizations, conducted missionary work in towns and villages, and so forth. Apart from the cultural protocol that we had to follow in collaboration with local authorities in these many cities, this newest project was the first time that I had to seek IRB approval in the DRC.

Honestly, I did not know how to get it; I did not know where to go or whom to contact. On top of this set of new obstacles, there was a major political obstacle; the fieldwork had to be conducted in a mining city where strategic minerals were exploited. For those not familiar with the DRC, the country's mining sector is a strategic sector with serious political, security, and economic ramifications. So, conducting research there requires a high level of tact. Nevertheless, I had no choice other than finding a way to gain access to the territory as a scholar first and Congolese second, and to get the project IRB approved. On the ground, I had to use all of the possible political and administrative strategies, including tapping into my personal social capital to do the work. Of course, not everything was recorded and made known to the academic administrators. In this process, I was also blessed to have had a supervisor who had done intensive research in similar environments across the world. So, every now and then I discussed my work with him on the phone and together we developed strategies. This of course helped to maintain a level of accountability and academic integrity while dealing with the caprices of international fieldwork.

## Learning the Unbalance

Sometimes going through life does not give us the time to reflect on the meanings, the lessons, and the way we are changed by these many, often unpredictable, experiences. My experience in academia and international research is one of those life experiences that I continue to cherish and embrace with joy, frustration, questions, and gratitude. One thing that I prepared myself for mentally, when I left the DRC in the spring of 2010, was an understanding that the years ahead would be of cultural, social, and emotional unbalances. These were important, unavoidable steps. Though my nature was still that of embracing new challenges and trying new things all the time, I did not know what to expect from these new experiences as I was jumped into the world.

## Language Unbalance

I know that I mentioned the language obstacle earlier, but I think it is important to put this conversation into the context of societal change and the academic environment. I moved to Cape Town South Africa with limited knowledge of English, with a clear vision to pursue education. Everything that took place had to align with that vision and strategy. The first obvious activity was to enroll in language schools. I went on to register at two colleges at the same time. The first college had a Congolese professor as the main English instructor. My cohort was mainly composed of many francophone students, too. The ambiance in the classroom was very friendly and the level of teaching was acceptable. However, this course was a mixture of English 101 and a bit of advanced English, directed toward social conversation and having little to do with academic training. I was excited to be part of this group, but I felt that it was not enough to prepare for the university. I managed to master some basics and was able to communicate in stores and public, but that was it. Thinking in French and sometimes in Swahili and speaking in English was a challenging experience. At times I felt like I was unable to communicate and could only enjoy the very basics of life and communication.

My life became conditioned by my limited ability to speak English, making social integration was extremely difficult. During such a time, I could, at least, rely on my weekly church services for interactions with many people, as well as for learning some new vocabulary. I remember one day I went to a store in downtown Cape Town during the winter season. It was extremely cold and rainy, and almost everyone was wearing gloves. Upon entering the store, which was owned by Pakistanis, I heard people talking about the cold and why we should protect ourselves. I then enthusiastically decided to jump into the conversation and agreed with them. I went on to mention that I regretted not wearing my gloves (but using a French word, *gants*, which sounds like *gun* in English). At the time, I thought that the word was right in place and that everyone would understand what I meant. I was wrong, but innocent and ignorant of the feeling of uneasiness I had just created in that little store. We kept talking and I went home. It was only when I was reflecting on my day and decided to double-check the dictionary that I realized how I had misused the word. This scenario is one of the many examples of which I knew the right word in French but could not communicate it in English, hence, limiting my social interaction with my community.

While studying English in this college, I enrolled into another, more specialized academic college that prepared students for universities. The instructors where a white couple, one South African and the other British; each had a very strong, distinct accent of their own. They were such an amazing couple who tried to connect to both South African culture and society while teaching academic English. Here again my challenge was that I was dealing with scholars who each had their own accents, wordings, and meanings of things; I found myself a Congolese man in a language tsunami. I quickly learned to understand that an accent was not a determinant of knowledge of a language. I had to stop trying to fit in and speak like the others (even though I still tried while alone). In time, my confidence grew, since my interlocutors could hear me. I kept, however, reminding myself over the years while in academia that language is solely a means of communication and not a determinant of knowledge. Throughout my years of international research and partnerships, I also came to realize that we live in a cosmopolitan world where our differences are our strengths. The same view should be promoted in academic spaces, where conversations of diversity and inclusion are considered taboo in some established institutions, many of which dream of nothing other than to maintain memories of past glories, which were anchored in the narrative of "us versus them."

## Cultural Imbalance

The sensitivity of cultural awareness was at the core of my heart and mind. However, most of the social and cultural practices that I had yet to be exposed to, unfortunately, instead of them being complementary to what I knew from my upbringing, were either somewhat opposing or entirely new. This was as true across academic practice as well as nonacademic. For instance, in classrooms and meetings in my home culture, one cannot talk without being given the official space and authorization to speak when gathering with those who are senior in age. In my new academic spaces, this was not the case. Yes, there was still as sense of protocol to be followed, but most of the time my colleagues would jump into the conversations and debate without waiting for the word to be given. During my few first meetings in this cultural environment, I felt like people were rude, unorganized, and sometimes hasty as they rushed through things. Little did I know that I appeared extremely moderate and weak due to my silence and nonaggressive approach. I then knew that I had to gear up and started jumping aggressively as well, even though it felt foreign to my culture and character. This was one of the most challenging experiences, since it touched the very core of who I really was. To avoid being completely denatured by this new culture, and to avoid watching my voice go into eternal silence, I made sure that I communicated my thoughts on what effective meetings look like and requested that I be given time to speak after everyone had aggressively jumped into the conversation. The tactic worked; however, the biggest lesson from these experiences was that I also had to be ready to adapt and survive the cultural imbalance if I wanted my voice heard and my knowledge shared.

## The Loud Silence

From the time I started this journey of international research and collaboration to the early days in my PhD program, I remained in a space of not being vocal or aggressive enough. One of the main reasons for such silence was that many, if not the majority, of those in my academic spaces had such Western views, that embracing another perspective was almost seen as a crime or a taboo. The danger of navigating an unknown territory and being exposed to a new truth in the realm of academic ideas and analyses was promoted over the opportunity to learn a non-Western perspective. I work in the discipline of international relations, with a broader training in global affairs and international development. These academic fields, in some instances, are primarily made up of Westerners, and the few contributions from the rest of the world are not valued. Theirs is unfortunately an invalid perception that does not stand ground to the light of historical facts. Regardless, this is the academic environment in which I, as a francophone and African scholar, must operate daily. I remember meeting with a very well-established scholar in the field back in 2015, who asked me why I was interested in international relations instead of African studies, or something else along those lines, as well as why I chose the US instead of Europe or any other francophone country. By the way, this scholar was also an immigrant from Europe who found himself teaching in the US. One could expand this conversation to revolve around many other issues such as race, the idea of supremacy, and so on, but that is not the purpose of this writing. That encounter made it very clear to me that my voice was not well received due to some established bias that had nothing to do with who I was as a person. Nevertheless, I persisted. Sound familiar? While it was easy to try to conform and disappear into the ocean of Westerners, and start speaking and writing like those who thought they had the right knowledge or owned the knowledge, I decided to keep my "other" perspective and defend it throughout. Like me, many other scholars find themselves in this type of situation and their voices become a loud silence. The reflection of transnational research and collaboration should be the reflection of multiple voices rather than that of one dominant voice.

## Picking the Microphone

History has shown that among the most difficult battles of humankind is the battle of freedom and voice. This battle continues to serve as a daily experience for scholars in many fields and regions across the world. Westerners struggle with it when jumping in the Global South, and those in the Global South struggle as well when jumping in the Global North. It becomes even more fascinating when some base their discussion on issues of ownership of knowledge. My experience and approach to this genuine struggle has always been that of making sure that my own platform is created and my own voice is out there. I believe such an approach to academic scholarship is critical if we want to be inclusive and fair when it comes to our intellectual curiosity. It is evident that the academic space is often seen as the most challenging space, where among faculty members, the excitement about knowledge production is always higher than the interest in gaining knowledge. This has been the reality for a very long time, and the direct consequence is that many voices went silent, some had to learn new knowledge, and others took the road of relearning their own knowledge.

## Suggested Readings

Adichie, C. N. (2009, Oct. 7). *The danger of a single story*. [Video]. TED. https://www.youtube.com/watch?v=D9Ihs241zeg

Addressing the power of stories and the disservice of a single story, Adichie provides a unique and powerful narrative as to why stories should always be heard within their multidimensionality. The place, culture, and language of the context are all integral part of the story, and therefore taking such approach in communication will avoid the imposition of views and perspective.

Holborow, M. (1999). *The politics of English*. Sage.

This book explains how English is used not only as means of communication but also a tool of power and influence in the hands of powerful nations. The dynamics of such reality reveals the nature of silence that the use of English imposes on other languages.

Kamwangamalu, N. M. (2003). Globalization of English, and language maintenance and shift in South Africa. *International Journal of the Sociology of Language*, *164*, 65–81.

The place of Indigenous languages in the production of knowledge seemed to be at risk in many African countries. This article not only highlights the multiplication and diversification of English, it also provides a strong foundation for understanding the battle of languages within Africa.

Louhiala-Salminen, L., & Kankaanranta, A. (2012). Language as an issue in international internal communication: English or local language? If English, what English? *Public Relations Review*, *38*(2), 262–269.

Louhiala and Kankaanranta reflect on the primacy of English language on local means of communication within organizational management. This is of greater value to the conversation of adaptation and use of appropriate language within a context of cross-border activities.

Samuelson, B. L., & Freedman, S. W. (2010). Language policy, multilingual education, and power in Rwanda. *Language Policy*, *9*(3), 191–215.

This case study of the historical view of language in Rwanda, the influence of colonial language, the challenges and opportunities that provide English in the new development agenda, and the role of Indigenous languages in local communities is a nice reflection of why language goes beyond education. It is a tool of nation-building and social cohesion.

## Discussion Questions

1. Do you think language can be a source of tension in education?

2. What do you think about the characterization of "Westerners" by the author?
3. If you were to go through this same or similar experience, what could have you done differently to come out from the "loud silence" that the author described?
4. Can you identify three risks and opportunities for jumping into the world?
5. Are there some benefits to using Indigenous language in education and transnational research collaboration? Please list three potential ones.
6. How well do you think the global system is prepared to embrace Indigenous knowledge into the mainstream educational curriculum?

## Author Biography

**Dr. Yvan Yenda Ilunga** is an assistant professor of political science at Central State University and serves as deputy director of the Joint Civil-Military Interaction (JCMI) Network. He holds a PhD in global affairs from Rutgers University. His research agenda broadly focuses on international relations, security, peace, and development; but more specifically, on questions related to humanitarian action, civil-military interactions, natural resources–based conflicts, peace operations, regional cooperation and security, and economic and social sustainability.

Chapter 9

# Chemistry Publication Ethics in China and the United States

## *Transdisciplinary Teaming in a Time of Change*

KATHRYN NORTHCUT

After 15 years of being moderately successful doing about the same thing over and over, I began to assume that the way I worked was the *only* way for me to conduct and publish research. My research was mostly text-based studies, limited ethnographic studies, and discourse analysis, most of the work involving just myself, a bunch of publications, and a keyboard. I did not foresee the situation changing. My place in the scholarly landscape is more about wings than roots, a unique trail of markers across a wide range of technical communication–related topics, from critical rhetorical studies to classical theory to empirical studies of social media. I was fortunate to be granted the promotions and the occasional award or honor, so there was no exigency for change.

But out of the blue in late 2018, I was contacted by research faculty in two far-flung departments on my medium-sized STEM-focused campus, and invited to join a team that was investigating research ethics among chemists in the United States and China. I'd never worked with chemists—I took a single chemistry class during my college career—and China was not on my academic bucket list. They were seeking an education expert, and I decided to consider this as an invitation to a Burkean parlor, and I'd happily dip an oar.

As I jumped at the opportunity for an unending conversation (again, invoking Kenneth Burke) with these researchers, I intuited what the project would look like and where it would take us. I thought maybe we would present ethics workshops to deploy with roughly equivalent groups in China and the United States Then we could compare results, follow up with focus groups or discourse analysis, and report it all in a single fabulous publication about how to teach ethics—which fit nicely with my (then) role as Institutional Review Board (IRB) chair. I quietly hoped that a second publication would materialize as a chapter in a book about compliance communication in a larger sense, not limited to science or to publication ethics. Our team's IRB applications swooshed off, primary investigators were certified, surveys were written.

The first thing I was happily wrong about was that minimal publication would result from our work. I'd long known the adage that any amount of research should generate a three-part payoff, like some holy trinity: presentation, poster, publication (or some equivalent), but I had never really been successful at maintaining that scale of productivity. Before I knew it, the science ethics research team members were generating data-collection instruments, pilot study data, reports, white papers, and networking opportunities—and I was racking up publications faster than usual. I requested my ORCID, a scientific tracking number that would forever connect my name and my publication record—in some journals, anyway. My National Science Foundation (NSF) biosketch was always updated. For the first time, I understood how scientists generate a dozen publications a year through collaborative hard work of teams, the ability to pay for travel, and creative ways of asking questions both within and across disciplines.

## Proposal Writing: Should It Really Be This Hard?

What I hadn't realized was how quickly research findings could be generated even when we were ostensibly *failing* to meet our prime objective. Failure in this case meant that our group had been unable to get past our internal campus review for a large NSF grant. We ended up placing third in a three-team competition for the privilege of submitting a proposal. It was clear all along that we were going to be underdogs, because the team that got the green light had submitted to this exact program at NSF at least once previously, although if points were given for being the fresh-faced new kids on the block, we might have won. I alone, among the team members, had been on campus for any real length of time. Among the things I'd done on campus was to develop proposal-writing courses for undergraduates and graduate students, which I taught for several years. The courses were in high demand and I received excellent reviews—some of the students notified and thanked me when they landed fellowships, scholarships, or grants. Again, incorrectly, I thought that those years of teaching grant-writing might give us some sort of edge.

Our internal review failure in this case wasn't upsetting; we had been working together a short time, and we had spent only a few hours on the draft. That writing had served the larger purpose of gathering our thoughts to approach more collaborators to pitch the project. By the time we knew that we would *not* be submitting to NSF, we were building our network anyway, and already becoming aware of alternative funding sources to expand the research; and we would be doing research about ethical knowledge of chemists, not primarily giving workshops.

By the time a year had elapsed, our team's efforts had borne fruit, having accomplished the following:

- generated tens of thousands of dollars of corporate and private gifts supporting the project
- built an international network of collaborators
- collected data from four different research instruments on empirical knowledge of science publication ethics
- reported the results of one survey in a regional science conference
- logged several communications with NSF personnel
- hosted two collaborators on campus
- cohosted (with a campus research center) two guest speakers with a connection to our ethics project
- hired a graduate student and an undergraduate student to help generate more information, and develop more publications

To be fair, some of those bulleted accomplishments had very little to do with me. But I had started to think of the team in the collective: individual contributions yielded team products. For the first time ever, I'd contributed to a paper that I didn't personally present at a conference. I was wholly dependent on teammates, and they on me. Our team meetings lasted hours. We went canoeing. More to the point, I expected them to use my work product as their own, and they expected me to use theirs. Our documents couldn't be divided into sections based on who wrote what; our plans were collaborative—guided by the group. My suggestion to do a multiple-choice test had been executed, but the test had already been revised based on two team members' data collection trips in China, and redeployed locally three times in a new iteration. We were starting to work as a unit, to row (intellectually) in synchronized movements. On one fine crisp day in October, two of us met with an NSF colleague in Missouri while two others presented our paper at a conference in Kansas. The whole really was greater than the sum of parts.

Prior to this experience, I'd collaborated with teams writing large NSF research grant proposals, and I had been at least peripherally involved in grants that would have yielded over one billion dollars. None of them had materialized. I had contributed to teams that received modest funding awards both externally and internally, but had never imagined becoming part of a research-producing machine like we had become. As we began to shift our focus from gathering survey data, moving toward the development and evaluation of teaching scenarios, we committed to wrapping up the first stages of our process with four publications (one high-value journal article and three conference proceedings).

By the start of our second year of work, we focused intently on our next shot at the annual internal competition for the NSF grant. Once again, we expected at least three teams to be in competition, as neither of us had yet been funded. After a pep talk from an NSF rotator from our campus, we were as ready as we would ever be to strategically submit the letter of intent (LOI) for the internal competition. We'd been advised to "downplay the budget" and to "be explicit about how the project filled the exact gap described in the program guidelines," complementing useful takeaways from extensive conversations with the NSF program manager.

## Transnational Study of Science Research Ethics

The global challenge of science research ethics is not new, but also not close to being solved. By definition and in practice, not all cultures have the same value system regarding intellectual property, plagiarism, reuse of material, requirements for authorship, and a host of other specific guidelines related to scientific research. My interest in these topics, from IRB compliance to publication ethics, quickly pivoted toward chemistry when I began collaborating with chemists on my campus. Intellectual property is just as culturally bound a value system as marriage or religion, and chemistry as a discipline is an excellent research case. With headlines featuring Chinese experiments using the gene-manipulation technology CRISPR, my collaborators and I have had plenty of exemplars for science ethics conflicts that might be rooted in cultural dimensions like national origin or region (West vs. East). And our team was tackling a complicated series of questions:

- Can science ethics be universalized without being culturally imperialist?
- Are professional ethical guidelines generalizable globally?
- Does instruction change declarative knowledge about science ethics?
- Does the method of instruction affect ethical decision-making among scientists?
- Does knowledge about science ethics change behavior of scientists facing ethical challenges?

Our team had been investigating questions of transnational research ethics, focusing on publication ethics, and we also exemplified the diversity of transcultural research collaborations. Although the composition of the team changed continually, a snapshot of the configuration during 2019 gives a sense of the diversity represented. Each PI claimed a different first nationality; each of us held a different country's passport. None of us were physically born on US soil. We all were employed at a public, state-supported university in the Midwestern United States, and each one of us had a different job title. Within our core group of four, three of us identified as female, with our senior scientist being a white male (who engaged with our critical commentary on topics ranging from *Queer Eye* to underrepresentation of women in science). Some of our work was undertaken in China, and some in the United States; having a native Chinese-speaking member made the research far more feasible, ethical, and comfortable. And when we added student assistants to our team, we increased our diversity with one male graduate student from abroad and one male undergraduate student from the US.

An ethical dilemma we faced demonstrates challenges academic researchers encounter, that researchers in industry may be able to handle much differently. The American Chemical Society and the American Psychological Association, both of which are examples of professional organizations who publish best practices for research, articulate guidelines that clearly state that unless someone contributes *substantially* to the research or manuscript, they cannot be listed as authors. Therefore, we chose to financially compensate a collaborator for some technical contributions in lieu of authorship; their contributions were somewhat minor, came late in the project design cycle, and did not take the amount of time that the authors had contributed. In fact, we soon learned that our own institutional policies prevented us from paying *any* faculty researcher to assist with professional services, and our institutional officials recommended that we instead give authorship credit to researchers who performed useful services. However, the work simply did not rise to the level of "substantial contribution" to the manuscripts. Therefore, the institutional guidelines directly conflicted with the professional organization/publication guidelines. What's a transdisciplinary research team to do? Turns out we couldn't pay our technical consultant, who responded graciously; we immediately involved them in developing another manuscript for subsequent publication, on which their role was expanded to merit authorship.

We also hit some major challenges in the differences between our own research styles of human-subjects research. I've always been advised not to identify the names of universities from which I gather data; that was not the consensus on all publications—to no real harm except my ego. As the IRB chair, I had some pretty rigid expectations about research protocols—and they weren't always shared by the entire team. More on that later.

We also encountered challenges in engaging and compensating collaborators in China, due to US funding agency policies, and we needed to be keenly aware of security concerns related to "Thousand Talents" programs and similar Chinese recruiting policies alleged to exploit US-funded programs and investigators. In fact, we wrote up a paragraph to include in our NSF submission acknowledging the existence of the Thousand Talents Plan, pledging to steer clear of such programs.

## Ramping up to High Gear: October Surprises

Transnational research is expensive; for example, funding for travel is quickly exhausted. US national funding agencies such as NSF don't financially support Chinese nationals, meaning that the bulk of our work has had to fall on those being compensated—the researchers stateside, even though we depend on our Chinese colleagues for access, introductions, and structure for visits there.

Gearing up to our second attempt at an NSF submission to fund our research, we had flown along in high gear for about a year. We developed four different versions of a publication ethics survey, took one of them to China where it was given as a pre- and post-test to over 100 Chinese graduate students, generated two publications, and secured two small grants for the work. We hosted speakers, arranged teleconferences, and otherwise worked mightily to be prepared as a science-ethics tour de force for the next round. Having lost our first try at the NSF grant, we were working backward from the previous year's deadlines. Aware that the NSF submission deadline was February 24, we planned a very busy editing and revising schedule for January. And having been rushed to meet the internal LOI deadline of December 14 last year, we scheduled a large amount of meeting and writing time between October 30 and December 7, so that this time, we would be finished *before* the date we had begun the LOI the last time.

The week before Halloween, on October 27 to be exact, I sent an email to an associate dean, who quickly responded by email that the deadline for the internal LOI was October 28! That date was published on a website, but if it had been announced, we had all missed it. I immediately forwarded the AD's message to the team. We had less than 24 hours; we hadn't started the major research push, and one of the team had just returned from the trip to NSF where she met with the NSF program director for this grant. I felt crushed. For a moment I knew that the most seasoned researcher on our team and I could stay up all night and generate an LOI, but I also knew that with two other teams vying for one university submission, our chances were slim, and a lot of work—hundreds of hours worth—had just been wasted. I still feel a bit nauseated thinking about it.

Quickly, in a flurry of emails and text messages, the team agreed that we should write to the on-campus senior research administrator who set and enforced the internal deadlines, to urgently request a month's extension to the deadline.

We never heard back from them. Instead, we received email from a staffer in the research office, and their message indicated that because no LOIs were received, they were putting it up for "first-come, first-served" status, and our email had given us dibs. We had the green light to develop our group's first NSF submission. We were elated—sort of.

Considering the amount of energy that we'd aimed at the internal competition, fueled by comments from two NSF contacts familiar with our situation, the exultation that we experienced on successfully navigating the internal funding hurdle was anticlimactic. I've previously observed the relatively unemotional way I perceive most of my engineer colleagues on campus responding both to grants won and lost. I've spoken with several researchers about the emotional impact of failing in high-stakes proposal writing. One of them told me you just get used to it; you write a lot of grants, if you're lucky you get good at it, and ultimately, you're too busy for reflection about feelings because there's so much work to do. However, in this case, our team's senior scientist admitted to being on the same sort of emotional roller coaster that I'd felt, albeit in an understated way. And we hadn't won the grant, of course; we had only won the right to apply for the grant. We didn't exactly pop a bottle of champagne—we didn't even have a meeting—but it was a good day.

## If You Want Something Done, Ask a Busy Person

In the midst of an overly busy Christmas break, our team leader looped me into an email discussion he was having with a scholar in the technical communication/rhetoric field, who currently held an administrative appointment in China—and their discussion involved inviting me to be a guest lecturer during the upcoming summer.

The email couldn't have come at a better time; I was overjoyed! We had collected data the previous summer, but NSF funding, if it were awarded at all, wouldn't be available for us to use in China for another year, meaning we stood to lose an entire data collection season. A guest lectureship would enable me, at a minimum, to meet contacts that could then be interviewed as soon as funding arrived, if invitation letters specified such.

We had several methods of data collection in the research design: direct interviews and surveys within China, and video conference interviews or focus groups that might involve Chinese researchers, but which would be conducted from the United States. The difference in these scenarios became important because of various permissions needed for international human-subjects research. The requirement of having a letter from the host, written in Chinese and English, explaining the exact nature of our visits to China had become very clear to us. We would need such a letter both to secure the travel visa in advance and for travel while in-country.

In some ways, the prospect of traveling to China was nerve-racking. I knew not one word of the language that didn't refer to food, cities are not my thing, and mass transportation is daunting. Dates that my colleagues would be in China were not absolute; neither was the time frame in which I might be teaching. I based my preliminary calendar on what happened the previous year, but soon I learned that the timing of that year was problematic. I learned that Beijing is smoggier in July than June, and the summer teaching sessions run from late May through mid-July. I'd hoped to travel to China right after Independence Day. That also cut it close with a major conference in my field I planned to attend later in July, partly to present our team's work. It was starting to look like, in order to make it to the conference that started July 19, my June family vacation plans (Glacier National Park) would have to go. Traveling to Beijing to collect data for the ethics study was important and compelling, and teaching in a smaller city was truly a once in a lifetime opportunity; if I couldn't do both, I'd have to choose. I vowed to be flexible, and hoped the glaciers wouldn't melt.

The lesson embodied in this experience was that close collaborations breed opportunities, but nothing is simple. Working with a colleague who had a long track record of teaching in China opened the door for me to imagine, for the first time, a summer lectureship in China. Productivity led to conflicting schedules in terms of when to write up, publish, present, and gather data, and we were of course all busy and struggling to find time to get everything done.

## Human-Subjects Research across the Globe

Collaborations are hard. Transdisciplinary international collaborations are really, *really* hard—but for all the reasons that make difficult research important. It's easy to do—or modestly expand on—what we already know, but much harder to identify or tackle a large knowledge gap. Add the need to study humans, and you've got a pretty complex situation. All

US academic and medical human-subjects research is subject to ethical oversight by a committee, usually referred to as an Institutional Review Board, by order of the Office of Human Research Protections.

Among the top-ten awkward conversations of my professional career is one in which I cautioned my collaborators that they shouldn't change an IRB-approved research instrument without approval. I was trying to explain why they shouldn't add questions that evaluated the course, the participants' knowledge, and their work habits. Perhaps the conversation was unnecessary, but we planned to administer the survey again, and I wanted them to not change it in the future—intentionally taking an extremely conservative approach in a highly uncertain research context. Eventually, I resorted to the regulatory language: "The study was approved to do X, but your questions changed so we are also now doing Y."

I tried to frame it in terms of participants' rights: "Our briefing script tells the participants we are studying X, but then we added Y. That's sort of . . . , you see?"

No, they literally did *not* see. I was sitting with academic colleagues in one of our long afternoon meetings. Late summer leaves rustled on trees and parklike lawns were visible through the windows. The precarity of studying human participants in a country without the IRB apparatus of the United States seemed concrete and tangible to me, but was completely invisible to others.

"But we made the study better, don't you agree? We noticed a problem and found a way to get more data that would help us learn what we want to know." It was very clear to my colleague that the additions to the study were wholly, purely positive, and the benefits to the study constituted the reason to make the change. The IRB approval was, to me, a barrier to changing the study; but to them, the study wasn't substantially changed—and even if it were, the benefits justified the change in the instrument and should not require any further approval.

I was reminded of Harris's *Rhetoric and Incommensurability* (2005), which first introduced me to the heady notion that my academic worldview might be incommunicable or inconceivable to people from other disciplines. If we truly cannot join a community of practice, and adhere to the shared norms of that community without leaving competing disciplinary ethical identities behind, what's the future of transdisciplinary work? That question is still current; cultural identity provides us with a framework that restricts our options even while it provides agency. At various times, my colleagues and I did seem to experience incommensurability over seemingly unimportant details—but that's one facet of working with people from another culture, regardless of the cultural dimension separating us in any given moment.

I asked my colleagues to defer to me when we write and deploy research instruments; they were grateful for my efforts, and we agreed to try to have a social scientist on site when deploying survey instruments to large groups of people. I knew that ethically, all our actions were defensible—we had studied the processes of researchers doing similar studies, we had local IRB approval, we were briefing our participants on the nature of the study in their native language, and we were not collecting any identifiable information. But my nature is to follow rules to the letter. Constitutionally, I'm just not the person who wants to apologize rather than ask permission when it comes to federal ethics requirements.

Although we sometimes think of research as a "right" of academic faculty, the process of trying to dot the *i*'s and cross the *t*'s to work ethically in China helped me conceive of research more as a privilege once I was aware of the enormity of practical, logistical, and cultural barriers, all of which are designed to prevent exploitation of humans. For various reasons—ranging from FERPA to local IRB guidelines—we may or may not be able to gain access to places, people, and data we need for research. I'd seen this situation in the past, in circumstances where an IRB could not approve research as the PIs requested. My experience chairing the IRB made our HSR application process much easier than it otherwise would have been, even though I couldn't approve my own study and I didn't control all aspects of data collection, curation, and analysis. Along with challenges rooted in our research training, other challenges we faced ranged from the stresses of collaborative development of major funding proposals to varying tolerance for uncertainty in making travel arrangements.

## The Ides of February, March, April . . .

By early 2020, the first hints of some weird SARS-like virus in China were circulating. Our research team, behind the closed doors of my colleague's sunny office, shared concerns about what these rumors would mean for our plans, which included annual trips to multiple locations in China to collect data. Not only had we already secured seed funds from a professional society to conduct interviews and attend conferences, we had funds from a private donor, and were about a week away from submitting the very large NSF grant proposal.

On February 15 of that year, I traded the relative calm of my faculty life for what was to become an academic administration hellscape, courtesy of SARS-CoV-2 and the disease we soon knew as COVID-19. As China started to look iffy, so did any trip to the US National Parks. By week four of my administrative gig, all the students except a handful had left campus in order to implement "social distancing" safety measures. At the end of week five, all the faculty and staff vacated our offices. By week six in mid-March, campus was closed and we were debating whether we were actually doing "online education" or "remote learning" (we settled on the latter), but mostly we were trying to keep operations going to avoid a crisis for every student and half the staff.

All plans—for research and teaching in China, for returning to classrooms in the US, and for normal life patterns, ground to a painful halt. At this writing, we are anticipating a rebuilding phase, in which team members hope to once again travel to China (and Hawaii, and many other locations), as we resume our networking and data-collection efforts and expand our professional repertoire.

COVID-related demands on our time turned three conference proceedings papers into one; conference cancelations further hindered our momentum. Some international conferences are still optimistically planned, but we're not expecting international conference travel to be approved by our university even if the conference *does* actually happen on schedule.

As the Fourth of July of 2020 fades into smoky, socially-distanced memory, we are left wondering about the kinds of commitments we can continue to make to our benefactors and our partners abroad. We know that the challenges of training a global workforce for chemistry research will continue to afford opportunities for our team to study the questions we have identified. I am thankful that COVID-19, while the greatest hindrance to our work, is an inconvenience, not a termination. We will continue to plan and to write, as our team navigates a shifting set of options and our priorities are necessarily reordered. Because the way we learned to work can no longer be the *only* way to work, and our academic work will continue despite disruptions we would never have predicted.

## Suggested Readings

Agboka, G. Y. (2014). Decolonial methodologies: Social justice perspectives in intercultural technical communication research. *Journal of Technical Writing and Communication*, *44*(3), 297–327.

To what degree are social science researchers responsible for equity in their transcultural research endeavors? Consider this argument and how it might alter transnational data collection activities.

Bankert, E. B. Gordon, Hurley, E., and Shriver, S. (2020). *Institutional Review Board: Management and function* (3rd ed.). Public Responsibility in Medicine and Research; Jones and Bartlett Publishers International.

Many researchers recognize that to study animals or do medical experiments, special permission and oversight are required. However, human subjects research regulations are inadvertently overlooked by many STEM researchers. For self-study in human subjects research history, philosophy, and regulations, this book is considered a classic (and recently updated).

Emanuel, E., Gadsden, A., and Moore, S. (April 19, 2019). How the U.S. surrendered to China on scientific research. *The Wall Street Journal*, C3.

Although we tend to think of the US as a permanent world leader in research, an alternative view exists. This article provides food for thought as we evaluate our perspective on collaboration, competition, and the nationalistic aspect of federally funded research.

Kalichman, M. (2014). Rescuing responsible conduct of research (RCR) education. *Accountability in Research 21*(1), 68–83.

Although many of us take a research methods course, the concept of broader "responsible conduct of research" (RCR) programs may not be familiar. Consider Kalichman's arguments, and whether they apply to the practice of technical communication, from both academic and industry-focused perspectives.

United States Senate Committee on Homeland Security and Governmental Affairs. (2019, Nov. 18). *Threats to the U.S. research enterprise: China's talent recruitment plans.* https://www.hsgac.senate.gov/imo/media/doc/2019-11-18%20PSI%20Staff%20Report%20-%20China's%20Talent%20Recruitment%20Plans.pdf

International research creates complications that US scholars need to study before undertaking projects abroad. Even the first few pages of this report will help researchers understand the caution the government now takes when funding research involving China.

## Discussion Questions

1. What kinds of experiences should technical communication students be expected to gain in order to productively engage with international, transdisciplinary research or teaching?

2. Identify a country that you'd like to conduct education research in, and determine how you will gain permission for human subjects research on site. What requirements or guidelines for *international* human subjects research does your institution provide?

3. Should universities compensate or reward faculty for attempting trans/international research, or attracting large grants from funding agencies, knowing that the risks (of time lost, for example) are substantial? If so, how? If attempting such research is not fruitful, how can faculty compensate for the "lost" time? If the risks are too high, how can we ensure that transnational research will continue?

4. Considering the intersection of regional and disciplinary cultures, what kinds of research questions are most interesting to you? Which are problematic? Does comparing groups in two countries inevitably lead to stereotypes? How can conscientious, ethical research about ethical behavior be designed?

5. Technical communicators often excel at gleaning information from subject matter experts. On what sorts of topics should you become a subject matter expert in order to pursue your career goals? What constitutes or demonstrates mastery of a knowledge domain, as opposed to mere familiarity or exposure?

6. Consider the prospect of incommensurability in the context of the type of transnational or transcultural research you currently conduct,

or wish to do. Consult Harris's *Rhetoric and Incommensurability* (2005) for inspiration.

## Author Biography

**Dr. Kathryn Northcut** holds a PhD in technical communication from Texas Tech University and the CIP credential from Public Responsibility in Medicine and Research (PRiMR). She is coeditor of two Routledge edited volumes, *Designing Texts: Teaching Visual Communication* (with Eva Brumberger) and *Scientific Communication: Practices, Theories, and Pedagogies* (with Han Yu). She is currently serving as vice provost for Academic Support at Missouri University of Science and Technology (Missouri S&T), and is a professor in the Department of English and Technical Communication.

Chapter 10

# Mingled Threads

## *A Tapestry of Tales from a Complex Multinational Project*

Rosário Durão, Kyle Mattson, Marta Pacheco Pinto, Joana Moura, Ricardo López-Léon, and Anastasia Parianou

We are here to tell a story of the Visualizing Science and Technology across Cultures (VISTAC) research project. Told as first-person narratives, the tales overlap in many respects. Together, they give a glimpse of an entrancing experience that lasted for five and a half years, from mid-2013 to end of 2018. From the start, VISTAC was housed at New Mexico Institute of Mining and Technology (New Mexico Tech) in the United States, but involved people across the world—Republic of China, Democratic Republic of Congo, Greece, Japan, Mexico, Morocco, Portugal, Russia, Spain, Sweden, the United Kingdom, and the United States. The project began with promise, had ups and downs, like any other, and ebbed away, like a wave at low tide. Figure 10.1 is the timeline of VISTAC's flow.

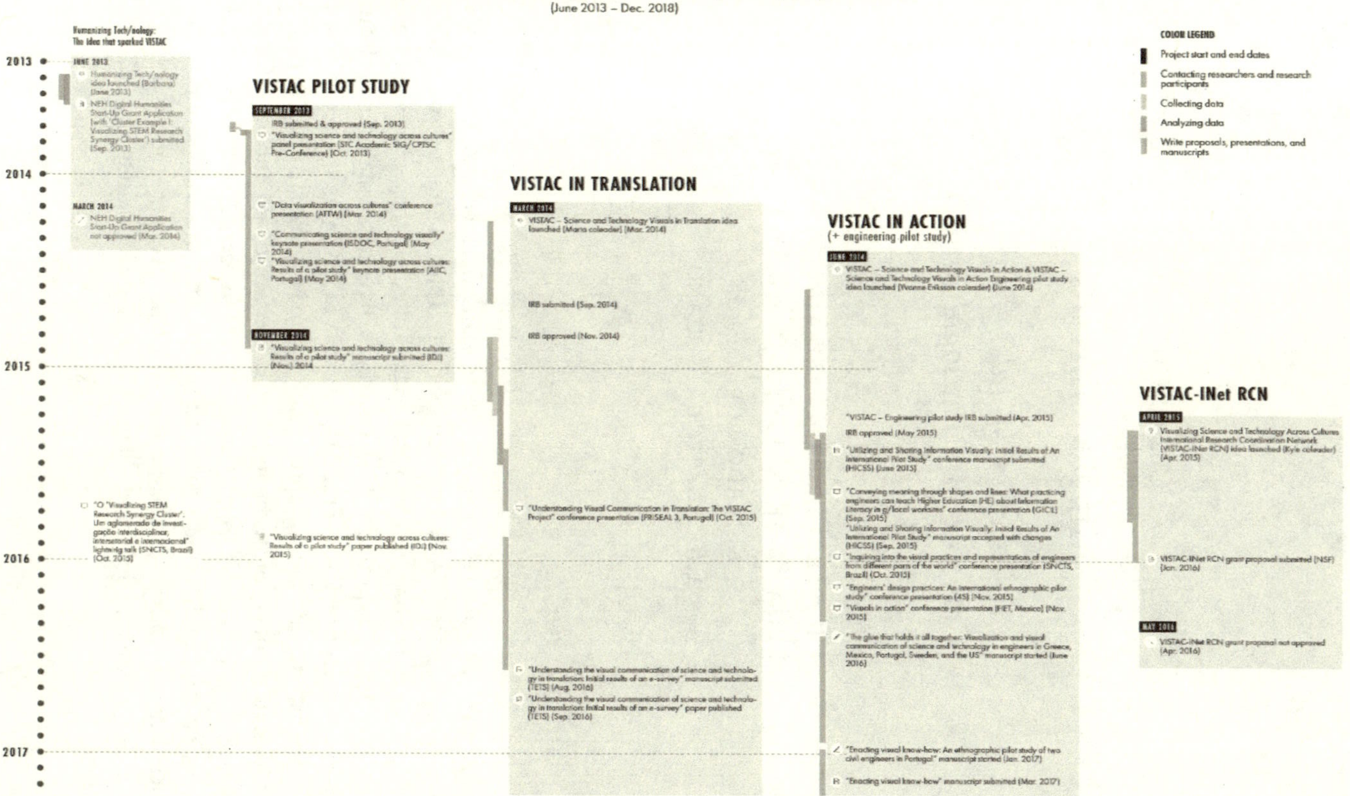


Figure 10.1. VISTAC project timeline.

## I

Rosário

June 2013 was a very good month for me, professionally. In the supportive atmosphere of my department, I joined a cross-campus discussion on promoting digital humanities (DH) at New Mexico Tech and other universities, decided on the research topic for my tenure, and created a research cluster for the DH grant proposal my department chair spearheaded. These all advanced my passion in investigating how data visualization, science and technology (S&T), and culture relate in S&T visuals.

In September, my department chair submitted the proposal to the National Endowment for the Humanities (NEH). Called "Humanizing Tech/nology: A Proposal to Integrate Humanities on a STEM-Campus and Beyond," it was to develop two research groups. One of them was the Visualizing STEM Research Synergy Cluster I would coordinate.

Excited about the project, I did not wait for a response from the NEH to start working on it. The day after we submitted the proposal, I sat down at my desk, looked at the large trees outside the windows, and set the wheels in motion. I began planning the Institutional Review Board (IRB) application for the "Visualizing Science and Technology across Cultures" (VISTAC) pilot study. When, eight months later, the NEH rejection email arrived, we had already done the pilot study, a panel presentation, and Marta Pacheco Pinto, Kristina Hennecke, Karen Balch, and I were analyzing the data from the pilot study (the pre- and posttest survey responses and freehand drawings of a science or technology concept) in greater detail for a paper we would send to the *Information Design Journal* in November 2014.

All the while, I was happily planning our meetings and sending out emails to the faculty and industry experts who had worked on the "Humanizing Tech/nology" proposal, and anyone else who wanted to participate in the VISTAC project.

This bustle does not mean that everyone stayed for long. With few exceptions, people came and went. You see, I followed Barbara's approach (Barbara was our department chair until she left in May 2014) of making everyone welcome, and publishing updates to the public on the Humanizing Tech/nology project website. This "flexibility" made leading the VISTAC project a fascinating exercise in managing flux.

It could also be quite frustrating, like the day when only one person showed up for a meeting in the large conference room we had set up for Skyping in—and that person was the student worker my department had generously assigned to the project. Take that meeting—I was sad, even angry when no one showed up. After a very long 15 minutes waiting for someone to pop in and watching the more and more sheepish look on the face of my assistant, I went to my office. It was a beautiful April afternoon, but I was oblivious of the sounds of students and birds outside. All I could think of was that I had emailed the notice to over 30 people, both at Tech and elsewhere, sent a reminder the day before, and a handful had said they would attend. So this didn't make sense . . . as the sun went down, however, "this" became an opportunity to rethink the "open-to-everyone" meetings. From then on, I held more focused (and more successful) meetings online.

Throughout the five years of this project, as I sat at my desk in my office at Tech or at home, as I worked in the US and in Portugal overlooking green spaces, streets, cities, or the ocean, and as the project branched into subprojects with people in different time zones and geographies, speaking different languages, and having access to different tools and technologies, making sure that everyone was on the same page was a daily concern.

To give you an idea of the types of information the project involved, let me list them:

- emails, PDFs, and videoconference recordings;
- Google Groups and shared mind-maps; Gantt charts and Google folders with lists of calls (paper and conference); a running list of bibliography; running discussion Google Docs; and several in-progress articles;
- different research management programs that I tried for sharing reference materials—though they never quite worked because of copyright concerns; and
- a shared online drive that I paid for myself because my university had trouble uploading videos to their storage space in a way that worked for our time pressures.

That drive was a wonderful solution for a way to store all the data we collected in the shadowing studies—for example, the two two-day shadowing studies I did in Portugal generated 65 photographs and 13 hours and 52 minutes of video from the shadowing, and three hours and 18 minutes of video from the reflective closing interviews. But the drive was slow to upload the files, crashed every now and then, and the researchers rarely went there because, although the data was well organized in folders and subfolders, there were so many files and folders that very few people, myself included, had the time to patiently sieve through it.

To make sure no one got lost in this maze, I sent out detailed emails every time we were working on something. In fact, email turned out to be the crux of the communication. It was my best friend.

Now that you know all about the overall project and how it was a challenge and an opportunity for growth, I will let Marta and Joana share the story of the VISTAC in Translation project.

## II

*Their eyes are fixed on the project: it emerged in 2014 under the leadership of Rosário, based at New Mexico Tech, who challenged Marta, stationed on the other side of the Atlantic, in Portugal, to embark on the study of how S&T visuals cross language barriers.*

### Marta

In one of the study rooms of the naturally lit library of the School of Arts and Humanities of Lisbon, Rosário and I designed a survey (something I did for the first time) to understand the overall practices and perceptions of translators, translation companies, and publishers when translating visuals in S&T documentation. The goal was to assess practitioners' awareness of the need of cross-cultural mediation when transferring S&T visuals into other languages, and geo-cultural and scientific systems. To join this squad on my side of the Atlantic, I invited my friend Joana, who was already working with me at the Centre for Comparative Studies, to bring along her expertise in mistranslation and visual culture.

### Joana and Marta

Our first step was to set up a multilingual team of coresearchers to conduct the survey in those languages that at the start of the project, according to Internet World Stats, were considered the top 10 internet languages: English, Chinese, Spanish, Japanese, Portuguese, German, Arabic, French, Russian, and Korean. The coresearchers were recruited via an online call for applications in September to October 2014. The call was circulated through our personal and professional contacts for researchers interested in assisting the project, specifically in terms of survey translation, dissemination, and data analysis. Potential coresearchers had to have working knowledge of English and preferably be native speakers of the languages in which we intended to distribute our survey. We asked each candidate to send us a biographical statement and a support letter explaining why they would like to collaborate. Against our expectations, perhaps because it was the first time we were recruiting multinational researchers on a volunteer basis and were thus skeptical about the call's potential to mobilize interest, we were overwhelmed by participants' responsiveness and eagerness to collaborate. All the languages were covered through the call except for French and Russian (for which we resorted to a more informal network, i.e., friends and family). Once the selection process was behind us, the project members translated the questionnaire from English into their corresponding mother tongues. It all seemed to be on the right track. After collecting the responses through FluidSurveys, our coreasearchers translated the (few) responses provided in their native languages into English; the data was then organized into an Excel spreadsheet for ease of analysis and comparison. The results of this survey were presented at two different stages—first the interim and then the final results—at international conferences in Portugal and led to a peer-reviewed article publication.

Before embarking upon the VISTAC project, we had been involved in collaborative and transdisciplinary projects, but never before had either of us worked with such a multilingual and geographically scattered team of people we had never met in person and whom we only contacted virtually. Despite our efforts to defy English monolingualism in S&T and scholarly research by translating the questionnaire into nine languages, some (Arabic, Japanese, Korean) did not even score a single response. It turned out that English was the preferred language of response, regardless of the place where the respondents were residing. This was frankly disappointing and made us feel that the work we had done for several months backstage in libraries and cafés in Lisbon had been in vain. It seemed that practitioners—not to mention the bulk of academics—give more credibility to English as a language of science, perhaps because of the widespread belief that English would also help them build an international reputation.

Eventually, we felt trapped by the responses we received: there was a mismatch between the purpose of the survey and the answers, which made us feel as though we had failed in our endeavor. Indeed, as we advanced with the translation of the questionnaire, difficulties raised our awareness about problems and limitations at the genesis of the survey itself. The results alone confirmed the need for designing a less quantitative and more descriptive and socioculturally oriented survey with fewer polar questions and more opportunities to elaborate on responses. Our questions were more concerned with setting borders between science and technology translation, when texts more often than not conflate both subject matters, and with confirming the relevance of visuals in these kinds of translations and determining what *we*, as researchers, knew to be common praxis—instead of finding out about the respondents' practices and perceptions and cultural constraints acting on their translations. This approach may have influenced and distorted respondents' reactions, when questions could have elicited more creative answers and made professionals visible by sharing their expertise in translating S&T visuals, for instance, by pinpointing strategies and exemplifying difficulties. Respondents were quite eager to share their practices and ask questions, including this insightful query from someone in Canada: "Why didn't the survey cover the question of HOW the visuals were translated (e.g., replace source-language wording if visual is clickable; add legend below visual if visual is not clickable; re-create visual)?" Despite the setbacks, this was most of all an important learning experience.

For both of us at that point in our trajectory, S&T translation of visuals was a new field of study. What made this VISTAC experience particularly rewarding was not only the cross-cultural exchange, but also the stepping out of our comfort zones to work on a project that allowed us to familiarize with current debates in S&T translation and in intercultural communication. Above all, it forced us to question our own visual literacy and learn more about teamwork in an utmost international context.

The VISTAC in Translation project ran for four years and suffered from some of its members' professional precariousness and the prioritization of personal commitments. It demanded self-discipline and persistence, which we would certainly not have attained without the enjoyment of working together on this challenging endeavor and supporting each other locally rather than struggling alone with the weight of tasks, responsibilities, doubts, fears, and frustrations, but also joys and accomplishments. This team effort paved the way for a discussion that hopefully other interested parties will further, relating to the cultural embeddedness of S&T visuals.

III

Rosário

While Joana, Marta, and I (they much more than I) were working on the VISTAC in Translation project, I spent most of my days and many nights working on another branch of the VISTAC pilot project: the VISTAC—Science and Technology Visuals in Action (VISTAC in Action) project.

I was fascinated by Latour, Law, and other authors doing science and technology studies (STS), and I had for some time wanted to investigate how scientists and engineers from different places and areas of specialization (1) pictured knowledge in their minds and (2) used visuals (and technology) in their everyday activities to express and shape that knowledge, and to *do* things in the world. So as we were writing the article on the VISTAC pilot project, I started thinking about the VISTAC in Action project.

Through a happy coincidence, I met Yvonne Eriksson from Mälardalen University in Sweden (she's an expert in information design and had done extensive research with industry). Because she had worked with engineers in Sweden, and had contacts in China, Brazil, Korea, and India that she could tap, we decided to start with an ethnographic pilot study of engineers.

Between June 2014 and May 2015, Anastasia, Karen, Kyle, Nabil El Hilali, Ricardo López-León, Tatiana Batova, Yvan Yenda Ilunga, Yvonne Eriksson, and I completed the IRB application, and got in touch with colleagues, students, friends, and family, and their networks to see if we could do a study with engineers. To everyone's surprise, this was very difficult.

IV

*Enter Ricardo and Anastasia.*

Ricardo

Excitement and concern were the mixed feelings that struck my body while reading Rosário's invitation to the VISTAC project. With a long-gone textile-manufacturing past, will the emerging industry in Aguascalientes be sufficient for this kind of study? What if I cannot find practitioners willing to participate? *¡No importa!* I did not want to lose this valuable chance and blindly accepted to participate.

As a visual literacy enthusiast, I have developed photographic activities for student-designers to promote critical thinking skills. If visualization was an essential aspect of scientific development, then innovation in science and technology could be promoted through training kids and youngsters in visual literacy and to train future scientists in developing countries. Since the Global Innovation Index was created in 2007, Mexico has never been among the first 50 places.

The main challenge for me was to find active practitioners involved with any product development or innovation processes to participate in the study. While dialing one number after the other from business lists, I noticed how committed secretaries are to their role as gatekeepers, since I could never speak to supposedly busy practitioners. Would you like to leave a message?

I confirmed my hesitation about the context. Mexico is still wild: the research-linked-to-practice culture is yet emerging. Hence, the industry has very tight security, and it is complicated to find someone that opens their company to a researcher to learn and capture inside processes.

As I shared my frustration with colleagues and friends, they also recognized the challenges and offered to help. Some of them recommended specific contacts through family or friends, and it was through them that I could finally begin to set up meetings to explain the study.

Meetings always took place in conference rooms where I had to broadly present how we do research and why it could be essential for understanding innovation processes. I often received skeptical looks, and inquiries circled an uncomfortable topic: "What's in it for me?" they seemed to ask through their euphemistic questions. The common good is probably not as integrated into the Mexican culture as it is in other countries. On the contrary, mistrust and playing it safe have had to be common attitudes due to insecurity and other social problems that this country has had to face in the last decade.

For that reason, practitioners took a long time and several conversations to trust me. Informal conversations were essential for developing trust, particularly when the audio-recording stopped. I found out that they were eager to tell their story. Focusing less on the process, they soon began mentioning how they began and all the struggles they faced to get where they are today. In my interviews with them, they proved themselves visionaries and appropriately disruptive game-changers.

Slowly they began revealing what they did and how they did it. Sometimes they rescheduled the appointment to see the process. "We are not doing anything interesting right now," some would say. When I could finally see the company's insides and processes, I was only allowed to record audio and take a few photographs of specific areas.

Since getting in took effort, I was relieved and grateful just to be there. I realized it was as difficult for them to let me in as it was for me to gain access. My patience paid off—a reliable relationship with industry! Perhaps these industry contacts will collaborate with me on future research projects or else recommend me to interested colleagues.

The VISTAC project also demanded a worldwide team effort. As such, discussing analysis and meetings were also challenging because of different schedules and world time-zones. Nevertheless, it was rewarding that despite language and culture barriers, findings were similar: visualizing is always a fundamental part of innovation processes, and each of the researchers could provide valuable insights into the project. Leadership from Rosário was excellent, and she always found a way to motivate the team, presenting herself always in a good mood while keeping us on schedule—and she would even agree to meet one-on-one to clarify doubts or discuss the project if it was necessary.

### Anastasia

Since I have lived and worked at the Ionian University on the mixed rural and urban Greek island of Corfu since 1999, my first thought when Rosário asked me to participate in VISTAC in Action was how few Corfiot professionals would or could take our survey. But then I remembered that my aesthetician, a Corfiot-born woman, knew at least one engineer at the Corfiot branch of the Technical Chamber of Greece. Because the well-maintained building (where the Chamber has its top-floor office) is nearby my home and comes with a view to the sea over large lime-and-black locust trees in the street below, I was very enthusiastic and made an appointment with the responsible engineer at the Technical Chamber of Greece to explain VISTAC and describe the goals of the project. I emphasized the scientific and international character of our project and that the next step would comprise shadowing studies and reflective interviews. She seemed to understand and promised to give the numerous copies of the questionnaires I made to her colleagues when she next met them at one of the regular Chamber meetings.

After the visit at the Technical Chamber of Greece, in early spring 2016, I waited for the person in charge to give me a call and hand me the filled-in questionnaires. How surprised I was when I did not hear back for some time! I finally had to call her to ask if any of her colleagues had completed the questionnaires. If I might hazard a guess, she may have forgotten to mention our project to her colleagues at all! I settled for asking her to fill in the form herself, but she never did. My frustration grew, so I turned to friends and asked them for their help. Finally, I met with two young engineers in Corfu who promised to fill in the questionnaire. When we met for coffee, they were reluctant to participate though I had explained VISTAC in Action; in the end I agreed to just read the questions to them and write down their spoken answers. I could never tell if they were too busy with other obligations or just indifferent. Happily, during March and April 2016, I received quite a few filled-in questionnaires from engineers in Athens—eight in total—thanks to some of my former students who counted engineers among their contacts.

I had even more trouble carrying out the project's shadowing studies and reflective interviews. While I met in person or by email correspondence with many professionals, quite a few of whom were willing to complete the questionnaires, the willingness of engineers to let me shadow them and ask questions during reflective interviews barely exceeded . . . zero! I had the impression that the few people I interviewed were not particularly interested in sharing information about their work, including visuals such as diagrams, sketches, photographs, or maps. Only one of the two engineers I met in Corfu—and who had filled in the questionnaire—was willing to meet for the interview. When we met at his office, he told me straight away how little time he had for me—he was in the process of moving to another office! I was fortunate to hold an extensive interview with a mechanical engineer in Athens. Though he was very stressed due to an effort to renew his employment contract at the time, he still answered all the questions I asked about the use of visuals in his work.

A real challenge for me was carrying out a sufficient number of interviews, even as Greece was undergoing harsh austerity measures of an economic crisis at the time of my interviews (February–April 2016). If I had the project to do over again, I would have preferred friends to strangers, if just to save time and nerves. I remain ever grateful to Rosário, the guardian angel of our VISTAC in Action adventure!

## V

### Rosário

On a crisp morning in Lisbon, in May 2015, I began a shadowing study that was unlike any other. I looked at the green door that looked like a stable door, on the other side of the street, wondering if it was the right one. After all, there was no number anywhere to be seen. My backpack was heavy with equipment to register Calvino's workday: computer, tablet, camcorder, cell phone, notepaper, batteries, cable, and my wallet. To my left, the cobbled sidewalks of Rua do Alecrim (Rosemary Street) slid down until they merged into a shimmering Tejo River.

Four men in T-shirts walked up the street, opened the green door, and disappeared behind it. A man in his early forties parked the car almost in front, took a laptop and some paperwork from the trunk, and also disappeared behind the door. I wondered if I should knock. But I decided not to. About ten minutes later, my friend walked up the street and smiled at me. She owned the small civil engineering company where Calvino worked. He was a project manager. The next half hour, as she, Calvino, and I sat outside a nearby café sipping strong black coffee, they told me about the project and what Calvino had planned that day. They were rehabilitating the first and second floors of a Pombaline building and converting it into a guest house.

When my friend left us and I went inside the building, it was like a war zone—dusty, noisy, cluttered. I moved carefully over tools, pipes, and broken tiles, among skeletal walls, boxes, scaffolds, sheets of drywall, and up old, worn-out marble stairs. Calvino introduced me to the friendly electricians, plasterers, painters, and other tradespeople working with him (those I had seen going into the building). I got out the camcorder, and began following Calvino around.

When the battery died, I transferred the recording to the computer, and switched to the tablet. When that battery died, I tried my phone and continued using it. It was lighter, charged faster, and was less conspicuous for Calvino and his crew.

Calvino spoke on one phone—even two phones at once—with his bosses, the client, suppliers, technicians, an inspector, city council people, and other stakeholders. I recorded him checking plaster along walls, sending and receiving messages and photos on his phone, making sketches on the walls to show what he wanted his collaborators to do, taking measurements with the AutoCAD app on his phone, and setting up his computer on a makeshift table. Everything was makeshift, actually. And, despite delays and other problems, everything worked!

I often felt like I was intruding, so I frequently moved a little back and recorded from farther away. At first, I video recorded everything—as if the lenses were extensions of my eyes. But when the narratives, actions, interactions, and contexts began to repeat themselves, I did fewer recordings and took more pictures of the sketches, documents tools, and contexts.

And in between all this, Calvino told me how Pombaline buildings were the first structures in Europe built to withstand earthquakes, and what it was like to work on these buildings: "We don't know what we'll find. And the work advances a little like the wind. We continue doing things. We know the final project, what we want the final result to be . . . [but] How do we get there? We constantly mold as the work progresses."

Over lunch in a tiny, bustling restaurant, he explained his new perception of the role visuals play in the workplace. And later, during the postshadowing interview, he expounded on this.

> My perception changed because I wasn't aware that I used visual means so much. I have to admit that I didn't have that perception. I am more conscious now that we use the visual component a lot to explain things. For instance, the scribbles we did on the walls—if I look at things from your point of view, it's different, but it was instinctive to pick up a piece of paper and try to explain to someone what I wanted. It was natural, normal, just a way of communicating. It's like talking, it's so natural that if you asked me, "Do you usually talk this way or that?" I probably do, but I really don't know. It's my way of being.

## VI

### Kyle

VISTAC began for me as a late-2014 email within a mix of ordinary correspondences—mostly about *connexions*, the online international professional communication journal Rosário and I once coedited. In the email, Rosário described another ongoing project. Seated at my converted desk, an old self-assembled pine table my wife inherited from her brother before he returned to Singapore in the late aughts, Rosário's email found me feeling like a kid looking up from the bottom of some well I had fallen into. Though I could make no sense of her multifaceted explanation at first, I was just glad to have been found in my midtenure jumble.

## THOSE INTERCULTURAL TRACES

As an early-tenure-track professor in a four-course-per-semester (colloquially 4-4) position, I parroted a confident "Yes!" like some echo of myself. *Yes to every single professional lead* was my mantra then, even as I was waking up to the US academy's perpetual version of a pyramid scheme—that every layer of accomplishment and privilege in the North American research community rests on the bedrock of those below. And each of us, no matter our situation in this scheme, says yes, no, or maybe from various unique points of opportunity, privilege (or lack thereof), and hope (or exasperation). Well, from where I sat at that old pine table, a chance to say yes to joining an international project seemed privileged enough. Besides, I was afraid to lose out.

Could I have picked up that habit of the flash-affirmative "Yes!" from my wife's Chinese Malaysian family in Singapore and Malaysia? Portrayed perhaps too often as a negative cultural trait or value, *kiasu*—a Hokkien (Southern Min) Chinese word lobbed about by Singaporeans, and some Malaysians, with a shrug and a Singlish (Singaporean English) *or Manglish* (Malaysian English), "Afraid to lose one!"—has its good aspects, too. There's that willingness to fight for family or for personal, if also family-contributive, outcomes. A win for me was a win for the family, right? But Singapore was nearly 15 years back when Rosário's email broke through, and I had returned to my "Minnesota Nice" ways over any *kiasu* leanings. I had difficulty saying "No" anyway, unless some overt social pressure caused me to dig in to stubbornness and delay the decision, and perhaps never get back to it, with a "That's something to think about" or some other "Minnesota Nice" alternative to having to commit. Add to this cross-cultural chess game a dose of post-PhD anxiety, and *Uff da!* No wonder I heard that old tenure-clock "ticktocking" away in Rosário's email.

## A WHEELHOUSE OF MY OWN

Perhaps I could pitch my competence to Team VISTAC as a wordsmith! Like so many US academics, I wanted to join international projects, even if my god-awful monolingualism moved me out of my depth among a cosmopolitan set of scholars with the multinational skills, presence, and access to move seamlessly among transnational research projects. Out of sync with the rest of the VISTAC team in that way (as well as in my self-perception), I bore a past of minimal success in an introductory language course—*or three!*—Spanish, Swedish, and Mandarin. Yet there I remained, stiflingly monolingual. Still, engaging with those beyond my own cultural sphere had still patterned my life, motivated my interests, and traced my heart and mind pre-VISTAC. Don't say no to yourself if others won't, right? While I would have to leave some aspects of VISTAC work to those with the right linguistic capacities, I would seek my value to our team through contributions I *could* make. Wordsmith it was!

Thus, I took on the wheelhouse of writer-editor, one of the main crafters, along with Rosário, of written deliverables for VISTAC in Action and Visualizing Science and Technology across Cultures International Research Coordination Network (VISTAC–INet RCN), the two VISTAC branch projects I had joined at her invitation. In this role, I helped our team articulate the complex intersections within, between, and among deliverables across a variety of shared online platforms. More about that later.

In practice, then, my role on VISTAC was *all over the place*. Two parts contributing-thinker and one part technical writer (not to mention three parts listener), I added to the shared conceptual stew while seeking answers from others to better clarify shared understanding and meaning, unify our message, and bring equilibrium to our texts, and, when possible, our team. The list below accounts for some of our—admittedly, largely unsuccessful—documents:

- one of the team's conference papers (never published)
- several team presentations (helped present a few)
- an article based on a study of engineers at work with a team of six VISTAC authors (never published)
- an NSF grant, VISTAC–INet RCN (never funded)
- Rosário's and my Portugal article (never published)

For Rosário and me, that Portugal article would become her late-tenure-track "leap-in-trap" from the data-rich depths of "The Glue that Holds It

All Together," a team-of-six article the team cared about more. In hindsight, I blame that admittedly bad decision on that same psychological pressure, the tenure-track clock "ticktocking" away for someone else: Rosário. Effectively making our decision as a team, we set aside "The Glue that Holds It All Together" (of all ironic titles).

## THE BUGBEAR OF INDUSTRY AND OTHER SITUATIONAL HANG-UPS

Like others at work on "The Glue" (part of that VISTAC in Action branch), I sought engineers as research participants near my US location, Greater Little Rock, Arkansas. In that effort, I had just one real good lead—a referral from a personal contact at a nearby nonprofit. That lead engineer soon dropped out, however, and although it was never clear to me what the real reason was for that decision, I suspect it fell through once I sought approval for taking onsite photographs and video—the old proprietary bugbear of industry! But whatever the reason for the abrupt end, I was left in a digital graveyard of dead "Request to Observe Engineer(s)" emails, ghosts of sent messages with few replies. My shallow list of nearby Arkansas contacts had grown cold.

The truth was that my contact list also included trained engineers in Singapore (e.g., one working for a US multinational, another in Singapore's mass rapid transit, and still another in heavy construction) and smallholder rubber-plantation owners in Malaysia—the latter with largely informal agricultural-engineering training (e.g., latex production and processing). Yet because those transnational contacts were all in my wife's family, a sense of propriety—"Minnesota Nice"?—kept me from leaning on those family ties, even when my personal goals could be viewed, somehow, as beneficial to my family. No, in my book those dinners, late-night coffees, and *teh gaos* (thick teas) in *kopitiams* (traditional coffee shops) were meant for the life and space of family in postcolonial time, not for some manufactured Western moment or agenda, no matter how I'd spin it. So as a cultural outsider given a place at the metaphorical family table of food-related hellos such as "你吃饭了吗?" ("Chi fan le ma?" = "Have you eaten?") and "Sudah makan?" (a similar meaning, but in Malay), I balanced *kiasu* with "Minnesota Nice" to leave those family connections alone, opting for what I perceived to be a more interculturally intuitive place, my VISTAC wheelhouse of writing, framing, "turn-of-phrasing."

## PRESENTATIONS ARE NOT PUBLICATIONS

Earlier, Rosário and I had joined others—Yvonne, Tatiana, Yvan Yenda, and Anastasia (one of this chapter's coauthors)—to prepare a presentation Rosário and I would deliver, here in the US, at the 2015 Georgia International Conference on Information Literacy (GICIL) in Savannah. There I was, "the tagalong kid," a few years down the tenure track, helping present timely transnational research for our team. Yet little did I know then that such success would not lead to "big ticket items" for our team—not during my time with VISTAC anyway. Overall, that late-VISTAC tendency to attend conferences and deliver presentations, rather than finish publications, meant we were often distracted from achieving larger goals. There would be no future articles, no funded grants within VISTAC in Action or VISTAC–INet RCN. Beyond the earlier VISTAC achievements, there would be no broad success beyond the goodwill, good memories, and active learning of working together as a globally distributed team (GDT). At the time I never thought of us as a (GDT), but the thought occurs to me now.

The GICIL presentation slides included photographs Rosário had snapped during firsthand ethnographic research in Portugal. Designed of vibrant pink and white, the slideshow showcased her flair for visually meaningful information design as it covered data sets from Team VISTAC's completed questionnaires. Organized by country, the questionnaires seemed to crisscross the world, but they were limited to our team members' own in-nation contacts. At the time, "Greece" (aka Anastasia) held the lead. She had collected eight completed questionnaires! Obtaining questionnaires from busy engineers (with their many commitments) meant we celebrated each other's successes.

## PLATFORM SCHMATFORM!

In its distributed work, Team VISTAC must have learned to use three or four online mind-mapping tools. First, there was MindMeister, an online application used to visualize bibliographies into relatable groups with nodes inside a mind map. Then there was Mendeley, today's Mendeley Cite, an online platform for sharing references and bringing them into our shared reference lists. And then there was Google Drive, Google Docs, and Skype—all mainstays of our team-distributed work at the time. Of course, some of these distributed platforms were rather complex and even overwhelming. The learning curve of one more online platform, with its bells and whistles, would become another obstacle to achieving our VISTAC shared goals, at least in those overburdened moments.

## ABOUT THAT NSF GRANT

VISTAC–INet RCN exhausted us from 2016 into 2017. While I had thought I would participate mainly in writing, revising, and editing the application, I soon *found myself* in correspondence and in-person meetings with the Sponsored Programs Grant administrator at my home institution, the University of Central Arkansas. Forecasting expected costs for a substantial institutional subaward and factoring them into the larger budget-forecast Rosário was working on at New Mexico Tech was daunting for someone who prefers words to numbers and paragraphs to budgets. I've since learned through casual correspondence that obtaining NSF funding is very difficult—a tiny miracle of sorts. Kudos to those who have successfully navigated that path! Were the budget approved, the subaward for my university might have been $92,661 of the total $499,999 VISTAC budget—more, of course, for New Mexico Tech, Rosário's institution. But when the NSF reviewers said "No" in an email, "All that work!" was our only thought.

## NO HULLABALOO HERE

Much like with the NSF grant, Rosário and my work fell short on a late-2017 article "Enacting Visual Know-how(s): An Ethnographic Pilot Study of Two Civil Engineers in Portugal." Despite the thousands of hours we spent gathering, analyzing, and explaining the collected ethnographic data on the role of visuals and visualization in the work of two civil engineers in Portugal, that publication faltered after three review cycles with *always conflicting reviews.* And there, again, was another email with some version of that word: *No.*

As one of the last VISTAC hangers-on in late 2017, I leaned into my own cultural disposition. I let the whole VISTAC team say "No" through collective lost interest over time. I wouldn't have to speak the truth that I suspect everyone on the team was thinking. Then the VISTAC edifice came down and faded away.

## VII

Rosário

As I was displaying this complex project in the timeline, I was struck by the breadth and beauty of the VISTAC project. I could go on and on about what went wrong and what did not happen, and the causes and consequences of this—complex wide-ranging project, busy schedules, let-the-data-speak methodology, no funding, some adverse reviewers, tenure committee pressure to publish in a technical communication journal, my tenure clock ticking (*importantíssimo!*), lost momentum, fading interest, and paralyzing guilt over this. All true.

But what the *visual* timeline made perfectly clear was the tangible and intangible things that went *well* and *did* happen: the research so many diverse and geographically distributed people did, the new people we all met and worked with, the mixed feelings of off-the-beaten-track ideas and doing innovative research, the collaborative writing efforts, the chance to share our findings in different places, the challenges and opportunities of each new moment.

For these and other things that went *well* and *did* happen, as much as for the mortifying things that did not, all of which marked me professionally and personally, I am, today, deeply grateful.

For my fellow travelers and weavers, a huge heartfelt thank you! We couldn't have done it without us.

## Suggested Readings

Groznya, E. (2013). Lost in translation. In H. Yu & G. Savage (Eds.), *Negotiating cultural encounters: Narrating intercultural engineering and technical communication* (pp. 81–101). Institute of Electrical and Electronics Engineers.

The author describes another's communication work for a transnationally dispersed team. Challenges emerged when cultural differences arose around communication styles and design choices.

Latour, B. (1987). *Science in action: How to follow scientists and engineers through society*. Harvard University Press.

The author describes science as a complex, hotly debated socio-human construction, and describes the role visuals play in the making of science. Latour arrived at this awareness after following scientists and engineers in their laboratories in France, the UK, and the US.

Longo, B. (1998). An approach for applying cultural study theory to technical writing. *Technical Communication Quarterly*, *7*(1), 53–73.

The author sees culture in professional communication work emerging from communication across organizations, not just within them. Researchers and practitioners must accommodate these landscapes.

Maylath, B., Vandepitte, S., Minacori, P., Isohella, S., Mousten, B., & Humbley, J. (2013). Managing complexity: A technical communication translation case study in multilateral international collaboration. *Technical Communication Quarterly*, *22*(1), 168–184.

The authors describe a trans-Atlantic communication project between faculty and students in higher education. Complexity is integral to the experience.

Tercedor-Sánchez, M. I., & Abadía-Molina, F. (2005). The role of images in the translation of technical and scientific texts. *Meta*, *50*(4).

The authors discuss several ways of analyzing images in source texts to judge whether the images need to be adapted to match the characteristics of the target audiences.

## Discussion Questions

1. What role does graphic communication, together with visualization as practice, play in the VISTAC project? How does the project timeline at the start of this chapter add value to the six authors' stories and perspectives? Is there anything missing from the timeline that you think should be added? Why or why not?
2. Consider how your future project work may depend on successful teamwork across borders (local, regional, and national). In what ways do borders alter team members' choices and inform changes to projects?

3. What do the authors mean by a globally distributed team (GTD)? How is the meaning of the term implied if not defined?
4. How does each story in this chapter work to reveal project successes and failures?
5. Why do you think the authors began the chapter title with the phrase "Mingled Threads"?
6. How do the stories work together as a "tapestry" to portray a shared team perspective of the project?
7. A key takeaway from this chapter is that, despite failures acknowledged in the stories, the VISTAC project succeeded in some respects. Do you agree? Why or why not?
8. Besides the reasons the authors named, what other reasons may have contributed to the successes and failures of the VISTAC project?
9. Have you ever worked on a challenging team project? In a page-length response, describe the problems and successes. How did the design of the project contribute, if at all, to project success or failure?
10. Based on lessons to be learned and aspects to be improved in the VISTAC project, put together a list of best practices for teams to achieve future success.

## Author Biographies

**Dr. Rosário Durão** is an associate professor of technical communication at New Mexico Tech, United States. Her background is in information design, translation, and Anglo-American studies. She teaches classes connected to visual communication, design, branding and social media, film, and international professional communication. Her current research is in the area of graphic and digital design. She founded and coedited *connexions: international professional communication journal* with Kyle Mattson. (rosariodurao.net)

**Dr. Kyle Mattson** is an associate professor of writing at the University of Central Arkansas, United States. Kyle's research focuses on stakeholder roles, rights, and responsibilities in designed deliverables for globally distributed and US-domestic sites of professional and technical communication practice. Despite Kyle's admitted struggle learning another language, he has not given up on his goal to speak Mandarin. Kyle coedited *connexions: international professional communication journal* with Rosário Durão.

**Dr. Ricardo López-Léon** is a lead researcher-lecturer at the Design Sciences Center in Autonomous University of Aguascalientes, Mexico. He has a PhD in art and sciences for design, in the area of applied aesthetics and design semiotics. His research focuses on visual literacy and design education.

**Dr. Joana Moura** is invited assistant professor in languages and translation at the Universidade Católica Portuguesa (UCP) in Lisbon, and a member of the *Literature and the Global Contemporary* project at the Research Centre for Communication and Culture, UCP. She collaborates in the project *Moving Bodies—Circulations, Narratives and Archives in Translation* at the Centre for Comparative Studies, University of Lisbon. Her research interests lie at the intersection between translation and literature.

**Dr. Marta Pacheco Pinto** is an assistant professor at the School of Arts and Humanities, University of Lisbon, and a researcher at the Centre for Comparative Studies, where she coordinates two projects: *Moving Bodies—Circulations, Narratives and Archives in Translation* and *Texts and Contexts of Portuguese Orientalism—International Congresses of Orientalists (1873–1973)* (funded by the Portuguese research council in 2016–2019). Her research interests include the history of translation, Portuguese Orientalism, and genetic translation studies.

**Dr. Anastasia Parianou** is a professor at the Ionian University, Corfu (Greece). Her teaching and research interests are translation studies, risk communication, specialized communication, major/minor languages, phraseologisms, and name studies. She has published three monographs on translation studies and a large number of articles and papers on translation. She is coeditor of the international translation journal *mTm*.

## Chapter 11

# Importing Lessons from Qatar

## *Toward a Research Ethic in Transnational and Intercultural TPC*

Nancy Small

The research presentation was being made by a group from another USAmerican international branch campus (IBC) in Doha's Education City, so I grabbed my notebook, closed my office door, and hustled through the spring heat to the building across campus. After showing my identification card to enter the building and finding the room, I settled in. The topic of the talk was outcomes of a research project on "the female majlis." In Gulf Arabian tradition, public concerns are discussed in the *majlis* or sitting place. Following the norm of gender segregation, men gather in a majlis to hash out issues regarding the neighborhood, town, city, or state. A men's majlis can be a circle of chairs outside, a formal space similar to a Western living room or den, a community center, or a lavish room in a government building. Almost nothing was written about the female majlis, so I was excited to learn about how Qatari women participate in the public sphere.

Eight or 10 folks were seated in the audience, most unfamiliar to me. I was happy to see my USAmerican colleague Matthew a few chairs away. He had extensive experience in the Gulf Arabian region, knew an impressive amount of Arabic for a native English speaker, and had raised daughters while living in the region, so I wasn't surprised to find him interested in the topic. One Qatari woman—identified as a member of the local population by her traditional black *shayla* (headscarf) and *abaya* (robe)—sat a few rows in front of me, and two Qatari female undergraduate students talked quietly in the back of the room. At the front of the room, a panel of Western white women prepared to present on their research.

The presenters discussed their project documenting the ways local women gathered and sought agency in Qatar's systems of communication and governance. Their research methods—interviews and observations—seemed sound, and the outcomes seemed to paint a positive picture of Qatari women's empowerment. The project was inclusive as well. Although the lead researchers were expatriate faculty members, they had enlisted Qatari undergraduate research partners from their branch campus. I was heartened by the direct participation of women from the community being studied. During the presentation, the professors called on the female students in the back of the room to comment and to share some of their experiences attending female majlis meetings as undergraduate researchers. The project seemed aligned with decolonizing methodologies, as it deeply engaged local participants and prioritized outcomes to serve local communities. It also seemed to be sensitively designed to respect Qatari culture and to resist viewing it from a deficiency perspective. In other words, Qatari women's agency was being considered on its own terms rather than in terms of what it "lacked" by Western standards.

As I listened with enthusiasm, Matthew bristled and grew progressively tenser. If her body language were any indication, the Qatari woman sitting a few rows up also seemed less than impressed. Even before the presentation ended, Matthew stood up and walked out of the room. I was confused by these reactions and wondered what critical perspective I was missing. During the question-and-answer period, the Qatari woman began asking questions. Turns out, this woman—whom I will call Dr. Qadha—was the *one* Qatari professor on the faculty in Education City at the time. Despite six campuses having been established there over the course of almost 15 years, few Qataris were employed at the IBCs. Branch campus faculties and staff were either imported from their USAmerican home campuses or hired from a pool of international professionals. Dr. Qadha was skeptical about several aspects of the study, but she was truly incensed by one: the Arabic name the researchers had assigned to the female majlis. The precise term they had chosen, which I don't repeat here, carried with it offensive connotations for women in Gulf Arabian culture, Dr. Qadha explained. She offered several other options that, she argued, would be more accurate and less offensive. The lead researchers listened to Dr. Qadha's assertions, but rather than address these concerns themselves, they asked the two female undergraduate researchers in the back of the room to defend the decision. The two young Qatari women stammered a bit and offered their rationale, but it was clear that they were uncomfortable defending the term to Dr. Qadha. She turned her gaze back on the professors and asked them why they didn't defend the term themselves, as the ones with power over the project. They politely declined to reconsider their assigned term for the female majlis, and although the tension in the room did not subside, the question-and-answer session turned to other topics.

What had I just witnessed? At the conclusion of the presentation, I walked back to my building, on the way stopping by Matthew's office to ask him about what the issue had been. He critiqued the insensitivity of the Western researchers in general and verified that the particular term they had chosen to defend was the reason he left so abruptly. Because of his knowledge of Arabic as well as his experience raising female children in the region, he understood and shared Dr. Qadha's offense. His early departure from the presentation had been intended to demonstrate disapproval of the researchers' decisions. I thanked him for sharing his insights and frustrations then left him to his work, still trying to wrap my head around the mixture of "better" and ultimately fraught—as well as potentially damaging—practices I had just witnessed play out. As an emerging researcher myself, motivated to learn and share more about this fascinating country and region, how could I avoid getting myself into similarly damaging intercultural and ethical entanglements?

## Wandering into a Transnational Scholarly Life

To continue my story, I need to back track a few years. In the summer of 2010, my partner, our three children, and I relocated to Qatar to work for an IBC of a USAmerican university in Education City, a collaboration of programs developed and sponsored by Qatar Foundation (QF), a generously funded extension of the Qatari government. Led by Her Highness Sheikha Moza, QF had invited branches of leading USAmerican campuses to Qatar so local students could remain near their families while pursuing a Western-style education. Arriving for what we thought would be a two-year term, my family and I began acclimatizing ourselves to a space that was, at once, both familiar and unfamiliar. We were grateful to drive on the right-hand side of the road but had to adjust to incredibly aggressive local drivers and the dread of three-lane roundabouts. English, in a glorious array of accents and vernaculars, was spoken almost everywhere in Qatar, but comprehension across those differences didn't always happen. We learned to laugh when these typically trivial misunderstandings occurred. Like the time when I asked a grocery clerk about where to find a "flyswatter" and he led me to a five-gallon jug of "fresh water." People were kind and approached such intercultural interactions with open hearts and enthusiasm for learning about difference.

As I began teaching technical communication and first year writing at the IBC, I learned that I had access to a generous faculty support fund for professional development, including graduate study. I jumped at the opportunity, gaining acceptance to a PhD program at a university back in the States, one with live online class sessions I would attend from 3:00 a.m. to 5:00 a.m., due to the time zone differences. The ensuing two years were a blur as I juggled parenthood, graduate school, and experimenting with teaching methods to more effectively meet my Qatari and international students' needs. During that time, I also had the transformative experience of writing the literature review described in this collection's introduction, an occurrence that catalyzed my search for advice on transnational research ethics. Our first two years in Qatar passed quickly and all five of us loved living and working there, so my partner and I renewed our contracts for another two years, then another two years after that.

Over these years, my colleagues and I attended international conferences where we presented talks over our teaching strategies and earnestly described the challenges of being faculty in the borderland spaces of an international branch campus. In our audience interactions, we learned to expect *the question*: "How do you reconcile that your campus is part of a contemporary colonialist project, that you are carrying on the damaging imperial practices of the past?" We had no good answer but knew it was a legitimate conundrum we had to continually face. The situation in Qatar was not straightforward: our IBC had been invited to Education City by the country's leaders and was fully funded by local money, and our sponsor, Qatar Foundation, specifically tasked us with delivering a replica of the home campus experience for the branch campus students. It was an "at-will" partnership. Just as we had been asked to create the program, we could also be asked to leave. Yet despite being invited guests, we *were* acting in colonizing ways because we were agents importing Western norms and assumptions, regardless of our endeavors to adapt our style and content to local norms.

During this period of growing awareness, I attended the Education City presentation on the female majlis shared above. I also began noticing other ethically questionable projects in which Western (also white, sometimes male but often female) academics would come to study the students at our branch campus. One semester, a team from a university located in the US asked a colleague and me if we would informally consult with them about a research project they were proposing as they sought grant funding from Qatar's generous national research fund. The USAmerican researchers would be in the host country for a weeklong visit and would appreciate our insights on their project design. We met with them and listened to their plan to study how Western education was helping to impact aspirations and elevate overall potential of Qatari female university students. We asked, "What's in it for the female students?" and were told, "We can help them see how education is improving their lives."

This story foregrounds how researchers with good intentions can naively romanticize a host culture and mistakenly rationalize the benefit of their work. The visiting scholars had assumed a portrait of Qatari women being uneducated and oppressed. They had not spent enough time in the host country to understand notions of "potential" and life "improvement" for the local community, so they defaulted to Western cause-and-effect assumptions about education and quality of life. In fact, as my students had shared with me, higher education could lead to significant social struggles for the females in Qatar and in the Gulf region. Education could risk creating family tensions if family members discouraged it or didn't like the field or major the student wanted to pursue. Outsider research projects like this one were motivated primarily by grant dollars, were insufficiently grounded in local relationships, and were inadequately focused on serving the host community. They perpetuated the long history of colonial projects that have already done so much damage to marginalized communities.

My close colleagues and I began having long and difficult discussions about our own projects and motivations, about the systems we were propagating, and about how to do more than proclaim our good intentions. These conversations felt circular, as we debated potential better practices but always ended up having to admit our very presence was problematic as a colonizing influence. We started research projects then scrapped them because ethical concerns stopped us in our tracks. Meanwhile, the clock was ticking, and I needed to design a study and write a dissertation. I had to learn to *perform* as a scholarly researcher. Out of this whirlwind of opportunity, transformation, and pressure, questions about research ethics sat staring me in the face.

I knew I needed to keep doing my own homework. First, I turned to Indigenous scholars including Margaret Kovach, Linda Tuhiwai Smith, and Shawn Wilson. They helped me begin to realize the damage Western researchers have done by objectifying colonized and marginalized peoples as "others" to be studied, and I knew that I didn't want to be part of perpetuating that harmful and shameful process. Indigenous and decolonizing methodologies were of great inspiration, but these authors were often speaking primarily to other Indigenous and marginalized groups. Who was I, as a white woman with a graduate degree and economic privilege, to blithely appropriate their work as a means of assuaging my anxiety and guilt? I sat with their lessons but sought out more. I wanted someone to tell me the *right* way—or at least *better ways*—to design, conduct, and write about research, particularly in transnational spaces.

Next, I turned to Google Scholar. With all the transnational and intercultural projects in technical communication and other neighboring fields, surely I could find guidance on research ethics for border-crossing projects. In fact, I did find guidance, but it was dispersed, scattershot, always partial and particular. Articles and books on participatory design and other inclusive methods advised me to center community needs, address injustice, and balance the academic drive for scholarly production (i.e., tenure-worthy work) with serving participants on their own terms. I knew I must write research outcomes in clear, accessible language and foreground practical application of outcomes for everyday folks. Publications about human-centered methods were useful but remained limited because they were abstract. I knew from my graduate classes that human subject research was "messy." Working across linguistic and cultural differences layered on more complexity and, with it, more risk of missteps ranging from clumsiness to harm.

Slowly, I began to gather threads of advice but found nothing as comprehensive as I had hoped. In other words, I found myself at a dead end for what I needed: the real-life stories, reflections, and lessons learned from scholars engaged in inquiry across cultural and national borders. The more I looked for these reflections and lessons learned, the more I noticed their absence. In the academic publishing world, it seemed editors and reviewers didn't want to know about struggles, restarts, and shifting plans. They only seemed to want the cleaned-up final story and the conclusions that can contribute to knowledge-making.

Because I couldn't find a cohesive statement of transnational and intercultural research ethics, I set out to develop one for myself. In 2014, my colleagues and I started pondering what a "transnational ethic of care" in cross-cultural teaching and research would look like. Once I had developed an argument for why we need a transnational research ethic and a reasonably cohesive foundation of guiding principles—a process lasting two or three years—I drafted my argument and framework into a scholarly article and sent it out to technical communication journals. The first journal rejected it within 24 hours: "We don't do theory." The second rejected it because, although the reviewers found it compelling and well written, it "didn't have a clear methods section." It was accepted into an edited collection, but then the editors wanted me to add a "mini-comprehensive lit review" to establish that no such ethic already existed. That literature review turned into a whole new 9,000-word chapter, and my original argument and framework for the ethic was dropped from the book. In fact, that's where this collection started, with the ultimate rejection of my work from other outlets.

As both a closing and an opening for this collection of stories from transnational projects then, I offer you the work I started developing myself because I couldn't find the guidance I sought: a first attempt at foundational principles for a transnational research ethic in technical and professional communication (TPC). These are a closing because they wrap up lessons that I have been learning along the way and that our coauthors have generously shared here. These principles are an opening, too, because they are a *proposed* starting place for a shared ethic. TPC practitioners, students, scholars, and teachers who work in complex multicultural and border-crossing spaces will have their own stories, principles, and questions for reflection to add. Developing a disciplinary statement of ethics is a massive community effort, and I hope to generate discussion toward that goal. The story must not end here.

## Toward an Ethic of Transnational and Intercultural Research

The remainder of this chapter proposes a set of four principles of ethical transnational and intercultural research. Each principle is state first, followed by questions to help unpack it, and further discussed using examples from the stories in this collection. If you are considering designing or participating in a project situated in any kind of borderland, where different cultures encounter each other, where knowledge-making will occur, and where decisions need to be made, then these principles, questions, and comments are meant to be useful to you. Because "cultures" form at many different levels—including organizational cultures and even microcultures for smaller groups—such work does not need to involve geographical borders, linguistic differences, or other typical "crossings." These principles and questions are intended to inspire more thorough critical thinking about project design overall as well as during each iterative and overlapping phase of your endeavor.

## Overarching Principle of Accountability

An ethical transnational/intercultural research project is human-centered and is based on relational accountability among the project hosts, participants, and stakeholders.

Some guiding questions follow:

- How will the project maintain a human-centered design prioritizing respect, rights, and dignity for all participants?
- What are the expectations and power dimensions of the relationships? How will the project foreground relationships rooted in reciprocity and integrity?
- What relationship-building experiences are reasonably possible? What relationships can be established within the limitations of the project resources?
- What are the implications of emotional investment in the project for the researchers, local stakeholders, and participants? Will relationships developed during the research process continue? If so, how? If not, how will they end?
- How are openness, humility, and gratitude demonstrated at every phase of the work, as a function of committing to and valuing relationships?

This umbrella principle is inspired and informed by two key writings. Rebecca Walton's "Supporting Human Dignity and Human Rights: A Call to Adopt the First Principle of Human-Centered Design" (2016), published in the *Journal of Technical Writing and Communication*, makes the compelling argument that ethical transnational and intercultural TPC research should be responsive to special challenges at each step of the research process. Her human-centeredness complements the emphasis on relational accountability in Shawn Wilson's *Research Is Ceremony* (2008). Wilson (Opaskwayak Cree) constructs his indigenist research paradigm on three pillars of ethical interaction: relationality, reciprocity, and respect. Relational accountability begins with ethical relationships among people but is a much broader perspective, encompassing acknowledgment and care of the complex web of interconnections among people, ancestors, nature, land, places, spaces, theories, facts, processes, and more. Human-centeredness and relational accountability go hand-in-hand, guiding the other principles as well as granular decision-making about a project's design, facilitation, and/or outcomes dissemination.

All chapters in this collection demonstrate the principles in this proposed ethic, but for the sake of space, I will highlight only a few stories to illustrate each one. Laura Pigozzi's chapter foregrounds human-centered design. Her primary concern throughout her project, from planning to writing, is the safety and, therefore, anonymity of her participants. Because some of them are unauthorized, she cannot risk providing any identifying details, and as a result, goes to special trouble to get Institutional Review Board (IRB) permission to waive the signature requirement for the study consent form. She is strategic about how she records study locations (through nondescript interior photos only) and participant interactions (audio only). Immigration status remains irrelevant to the study, and she destroys signup sheets straightaway. These moves—which also correspond to responsible preparation—are a result of Pigozzi's human-centric concern over the privacy and safety of participants and their families. The federal agency US Immigration and Customs Enforcement (ICE) raids occurring in the background made such protections a primary priority.

Pigozzi further demonstrates relational accountability through her preparation when she incorporates input from participants and from community leaders into her research design. As a result of their advice, she becomes more aware of the detailed ways she presents herself, she shifts from individual to group interviews, she prepares food for the participants, and she creates spaces where children can feel welcomed and engaged during the interviews. When Pigozzi finds out her grocery store gift cards have not functioned as intended, she adapts by getting permission to thank participants with cash instead. She is concerned about the trust established between herself and her community, so she approaches the change from gift card to cash not from a perspective of functionality but from a perspective of relational accountability and reciprocity. She prioritizes her relationship to the broader Latinx community and to her church parishes. Although she has been a long-standing participant in her local religious community, she knows she must guard the ethos associated with her positioning and that she must be sensitive to the privileges she brings to the community collaboration.

Sarah Beth Hopton, Rebecca Walton, and Linh Nguyen's chapter amplifies the importance of a human-centric design and relational accountability through their collaborative partnership. Hopton and Walton reflected early in their planning process about their own anticipated shortcomings in talking with Vietnamese participants about the rhetorical framing and communicative impacts of Agent Orange. Help with linguistic translation was imperative, but they purposefully sought out a partnership based in shared interests and commitments. Their decision to work with Nguyen as a cultural ambassador, someone who could both translate the language and interpret contextual elements, was a move of relational accountability to their Vietnamese participants, to Vietnam as their host country, to USAmerican veterans affected by Agent Orange, to the Vietnam War/American War, and to the knowledge-making process they were pursing. In these and other ways, Hopkins and Walton emphasized their commitment to deeper engagement with and growth through their project.

In their chapter, Hopton, Walton, and Nguyen each describe things they learned and how, as their relationship grew, they began to feel more accountable to each other. Some of these details may seem trivial, such as Nguyen's advice that a shirt needs a collar in order to demonstrate real professionalism. However, these small details—many of them nonverbal—are threads that weave together in how researchers and practitioners present themselves across difference. Could Hopkins and Walton complete their interviews without collared shirts? Most definitely. But would the lack of collars and maybe some other small, often unconscious details, potentially contribute to a poor overall cross-cultural impression? Likely so. The spirit in which Hopton and Walton were open and responsive to Nguyen's coaching illustrates another principle of this ethic, responsive adaptivity, but their motivation for being adaptable is their human-centric and relationally accountable mind-set.

Nguyen joins the USAmerican partnership already formed between Hopkins and Walton, and her reflection over experiencing actions based in integrity demonstrates how relationality grows in all directions via transnational collaborations. Nguyen's work translating and interpreting consent forms with Vietnamese participants expanded her understanding of confidentiality as culturally contextual. The consent process's emphasis on protection is human-centric but is also embedded in relation to USAmerican values, which assume that individuals have private, sensitive information potentially controversial to the community. As Nguyen talked with Vietnamese participants, she learned that privacy and consent can also be in relation to trust with regard to respecting emotional and physical trauma, at both the individual and community levels.

Yvan Yenda Ilunga's chapter expands the perspectives in this collection by telling stories of his experiences working to become a member of the scholarly community and the relationships between language and belonging. His narrative generously offers Western English speakers the opportunity to reflect over how English being the lingua franca of research marginalizes speakers of different languages. Ilunga's narrative invites reflection over how language differences might influence the human-centeredness of a project, when translation activities alter interactions and when interpretive assistance (or lack of it) shapes meaning. His stories reveal a web of relationality among language, access, agency, confidence, institutions, disciplines, and knowledge-making processes. In particular, when he requests his colleagues take note of and make space for different interaction styles, Ilunga shows us what it's like to be a human on the margins seeking out accountability from others. His endeavors to cross multiple boundaries in order to join an anglophile scholarly discipline linger as reminders to be more attentive to our colleagues' and participant's diverse skills and perspectives in relation to what we center in our work.

Ilunga brings up another excellent point in sharing the story about returning to his home community in the Democratic Republic of the Congo (DRC) to conduct research. One of his faculty reviewers asks how he can possibly maintain objectivity in an environment where he is an insider. This question points to another important aspect of relational accountability: one's position as an insider, outsider, or in-betweener. By setting his study in the DRC, Ilunga positions himself in a liminal space between the academic world and his home world. That positioning is one of strength, as he can serve as his own interpreter and he has the advantage of deep knowledge and experience with the context. However, it also presents new challenges in requiring him to perceive his old world through new perspectives. In reflecting over tensions created by his multiple accountabilities—to his home community, to himself, to his project, to his academic advisers, and to his disciplinary standards—Ilunga comes to question how or why a researcher should be required to leave his affective ties at the door as a means of demonstrating "rigor" in a project. His storytelling reveals the complex entanglements of relational accountability and brings us back to the researchers themselves as being "humans" that cannot be discounted from keeping our projects ethically "human-centric."

Finally, you may have noticed some cross-referencing among the authors. Emily Petersen and Breeanne Matheson wrote about different research experiences but were also collaborators on Matheson's India trip. Yvan Ilunga happened to meet Bernadette Longo on one of her trips to the DNC, but that chance meeting was social, not related to either of their positions or research at the time. Later, however, Yvan's and Bernadette's paths crossed again as their shared interests brought them into conversation. Another demonstration of interchapter relationality is in the readings we have suggested. Above, I rely heavily on previous work by Rebecca Walton in establishing this first principle of accountability, and in the introduction, Bernadette and I refer to other work by Breeanne Matheson and her collaborator, Cana Itchuaqiyaq. The point is that the relational networks extending out from transnational and intercultural projects are not just about researchers, participants, and local communities. These relational webs encompass other kinds of connections to colleagues, documents, scholarship, processes, systems, communities, and situations. All chapters here have elements of this overarching principle threaded throughout them. Acknowledging that human-centered relational accountability is imperative to ethical project design, yet tracing and maintaining it can prove incredibly complex. The remaining three principles are, on the surface, one step more directly applied because they are divided into phases: designing and planning, data gathering and analysis, and writing and reading. Because project design, whether in the classroom, field, or workplace, is iterative, these principles overlap and intertwine. Things are never as simple or linear as expected.

## Principle of Responsible Preparation

An ethical transnational/intercultural research project requires additional preparatory investment in deep learning about the research situation; the host population, stakeholders, and participants; and the self.

Some guiding questions follow:

- How will you learn about the community where your project will be hosted or located? What do you not know, and how will you learn it?
- What cultural, historical, economic, and other contexts shape the research situation?
- What notions of identity (ethnic/racial, regional, national, gender, ability, age, religion, immigration/residency/citizenry status, social/situational hierarchy, etc.) impact the study, and how do those notions affect the host context?
- What relationship building needs to *precede* the actual project? How might that process proceed? Do researcher and host community notions of relationships (e.g., how they are built, how they are nurtured, the expected duration) align, overlap, diverge?
- What are host community notions of dignity and respect, and how is the project designed to align with and practice those local notions?
- What is the nature of the relationship between the people and the land in the locations where you'll work? How do those relationships tie to notions of ownership, trauma, hierarchy, history, and memory?
- What are the rituals, symbols, and etiquette underpinning interaction in the host location?
- How do the selected theory and methodology shape the project in ways that are complementary to and/or in tension with host community norms?
- What cultural ambassador relationships are possible? What role(s) will those ambassadors fill? What status will they be given in the project?
- How long do you expect the project to last? Do you have a pipeline of resources to support the project if it lasts longer than you initially anticipate?

- Do the expected project risks, outcomes, and benefits reflect a strong relationship and authentic reciprocity between the researchers and the host stakeholders?

- What are your individual strengths, challenges, and limitations as someone coming into this community to participate in the project?

Some of these questions may seem more pragmatic, such as those related to resources, but they are crucial because they underpin how we address the first umbrella principle. Proceeding with a poorly funded project or one that is not likely to receive further funding impacts how project communities and participants are viewed. As previous chapters have acknowledged, a project with limited funds might turn out to be "parachute research" or "smash and grab." A project where a return trip is needed but not feasible may result in underdeveloped or inaccurate contextualization as outcomes are written and disseminated. Also, note that none of these lists of guiding questions can be "complete" because every project has its particularities.

All of our stories demonstrate thoughtful energies devoted toward this principle of responsible preparation, but here I highlight three. Emily Petersen's list of tasks ahead of her archival tours starkly illustrates her extensive work in service of this principle. Among her efforts, she read deeply, collaborated on a major literature review, communicated with Botswanan and South African archivists for over eighteen months, and met with a South African colleague to discuss trip planning. She had been self-reflecting about her identity, positionality, and privilege as a researcher, recognizing that being white, speaking English, and coming from a Western country were markers of colonial hauntings. Petersen even notes that she and her USAmerican collaborators were aware that their knowledge and planning had to be flawed, that they knew their information—no matter how meticulous—must be incomplete.

Petersen's narrative demonstrates that although preparation can be intense in the earliest phases of a project, a forward-thinking approach proves valuable throughout the entirety of the work. Visiting the Apartheid Museum in Johannesburg brought new dimensions to the readings she had been doing and added to the depth with which she could understand the context of their project. Preparation involves anticipating what will happen as well as what might *not* happen. Petersen and her colleagues tested out different applications for saving archival documents, realizing that output quality, file sizes, and even internet access would likely vary. With regard to the permissions required from the director of the Botswanan National Archives, Petersen knew that the signature might never come through. Even though that unfortunate situation happened in the midst of the trip, she knew some struggles would occur. In other words, she was prepared to understand her place in the transnational and intercultural systems within which she worked. She knew that she was not entitled to make demands. Her narrative illustrates that preparation is both action-based and part of a mind-set.

Breeanne Matheson also shares stories of her preparation for the trip to India to research the experiences of women in technical communication. She and her collaborator (Petersen) delved into scholarly texts, carefully formulated their interview protocols, and set up community partnerships in the host country. They also considered their positionalities in relation to their plans and participants. Similar to other chapters about research with human participants, Matheson includes mention of IRB review and approval. Just as Petersen's chapter on archival work met with unanticipated turns, Matheson's chapter describes how, despite the rigorous planning, their preparation had to be revised once on the ground in Chennai. Their accommodations had to be reestablished based on their space needs and their interview plans had to be amended. Once they had started gathering responses from Indian women, they realized their questions were not a good fit—they had to recalibrate and loosen their expectations about intercultural concordance between Indian and USAmerican women working as technical communicators. Matheson's story demonstrates both preparation and agile adaptation.

The last guiding question suggested under this principle of responsible preparation asks, "What are your own strengths, challenges, and limitations as someone coming into this community to participate in the project?" Matheson's chapter highlights a crucial but often diminished or erased implication of this question: transnational and intercultural projects are deeply embodied. Matheson, Petersen, Amaya, Hopton, Walton, and I refer to our identities as white, Western women. However, embodiment also affects health and feelings of safety. Matheson and others comment on the brutal exhaustion caused by jet lag that actually impacts basic functions as well as the kinds of quick and careful thinking that intercultural communication requires. Matheson graciously shares the story of their interview quality being impacted and the feelings of regret they had over feeling they had been rude. She also shares the story of witnessing a traumatic incident in her hotel and how it affected her for the remainder of the trip. Of course, no one can plan for situations like she experienced, but similar to Petersen's expectation that her preparations were inherently incomplete, transnational and intercultural projects should be crafted with the realities of our bodies factored in. While this sounds so simple and logical, the pressure to make every moment on the ground productive—as evidence of being accountable for funding and a responsible representative of one's organization—denies the realities of the situation. Here is a major tension between accountabilities to funding and institutions; to research collaborators, communities, and participants; and to ourselves.

Bea Amaya's travels through Papua New Guinea were certainly embodied as well as thoughtfully planned. Because her project involved a longer relocation to the country where her work was located, she had more substantive opportunities to learn about the local culture and social systems. Choosing to live in a PNG neighborhood and create relationships with the local folks through Friday movie nights allowed Amaya to deepen her contextual knowledge about the country beyond anything books and articles might tell her. Wearing her *bilium* (traditional bag) brought her insights into regional differences within the highly diverse country and opened her mind to material ways Papua New Guineans signaled and acknowledged belonging. Something about carrying that *bilium* and being associated with the Highlands may have helped her figure out the reason why her stakeholders reacted to the printed annual reports the way they did. She could not have planned to interpret that reaction, but her choice to live and socialize within the local community prepared her for her in-country trip in ways that living in an expatriate compound would have missed.

In good TPC fashion, Amaya also planned out the Annual General Report she was tasked with preparing for the stakeholders. She reviewed the previous reports' structures, content, and design, and she attended to its primary and secondary audiences' preferences for oral communication. Even if the reception of the material report provided a surprise, its contents, format, and design proved useful for the company meetings. Although I risk sounding presumptuous in saying so, I imagine that Amaya's pretending to enjoy a beer and a cigarette as a means of establishing belonging in her company's male ranks was mindful planning on her part. Because she had already experienced a career in the male-dominated world of multinational oil and gas companies, she would have anticipated needing to create her own space within the preexisting practices of the PNG company, rather than expecting them to come to her or a more neutral third-space to emerge.

A final note on something interwoven through the narrative of all of our authors, including Petersen, Matheson, and Amaya: a sense of openness and humility. Acknowledging the limits of secondary research, the short-sightedness of IRB approvals, and the likely overlooked contingencies in even meticulous itineraries is a form of responsible planning. Knowing that, in a transnational and intercultural project, we ultimately can*not* be fully prepared should incline us toward an open and agile mindset. That said, locating the time, money, and other resources to establish a transnational project's groundwork can be difficult and likely will not be enough to ensure smooth goimg. Even our best efforts learning about the host context and reflecting over positionality may not ensure an easily facilitated, ethical, and successful project. Our ultimate plan for struggle and surprise then brings us to the next principle of adaptivity.

### Principle of Responsive Adaptivity

An ethical transnational/intercultural research project is grounded in listening, is agile, and is responsive to the complexity of gathering and analyzing data across perspectives.

Some guiding questions follow:

- How will the project and its leaders/facilitators respond to the need for in situ agility?
- How will the project respond to challenges or tensions regarding data gathering methods?
- How will the project leaders and stakeholders react to unforeseen change, including time delays or unsuccessful relationship building?
- How will project leaders and stakeholders respond to potential opportunity or serendipity?
- Are the project leaders prepared for potentially drastic shifts in project plans and/or stakeholders? What elements of the project are "nonnegotiable"?
- How is the meaning-making process being adapted to include host stakeholders or participants? Is further adaptation possible?

The previous section, on the principle of responsible planning, gestures insistently toward the imperative to listen, to be agile, and to adapt. Transnational and intercultural projects, particularly when grounded in a universal principle of human-centeredness and relational accountability, must be sensitive to emerging needs and amendable to change throughout all phases of their cycles. The concept of an "agile attitude" comes from user-experience testing because the design of human-technology interfaces requires flexibility. Humans are wonderfully and woefully unpredictable. How we deal with those surprises says a lot about us as individuals, teams, organization, and a discipline.

Bernadette Longo's chapter on her telecommunications project in the DRC provided a case study of an agile attitude and adaptivity in action. Her narrative began in the disappointment of discovering her class's 2008 communication audit had not reached its intended community. Follow-up about the project's impact enacted relational accountability, and Longo's willingness to listen to and process the news that the work had not proved as relevant as they had hoped, in part, triggered her revision of the project. Over the course of three years, that website redesign evolved to exploring cell phone–based information transmission first for women's lending circles and then for miners and farmers. The emphasis shifted from social networks to business networks and from technical issues to pricing information, and the research questions grew progressively more foundational. I can imagine the collaboration might have felt like it was moving backward or in a meandering circle rather than forward. However, Longo's motivation remained consistent. She wanted to develop an authentic and reciprocal partnership that would help a group achieve its goals and would help her students learn about the real-life complications of information design.

In essence, her willingness to adapt and persistently invest in the project illustrates the human-centered and relational accountability to transnational and intercultural research. The project changed in response to the availability of her DRC collaborators' needs and challenges, yet she and her students (over the course of three years) persisted. Rather than frame it as a transactional give-and-take, Longo sought out situated relationships to allow her and her students opportunities to respond in meaningful ways. Longo's willingness to adapt and respond drew the project out to be three years in duration, yet it ended with no direct outcomes. Her resolution to learn from what would be labeled a "failed" project illustrates yet another adaptation. Instead of turning away from the experience, she chose to publish on her struggles. Her article, "RU There? Cell Phones, Participatory Design, and Intercultural Dialogue" (2014) was one of the few publications I could locate as I sought out stories and advice from experienced transnational researchers.

While Longo's chapter in this collection foregrounds relationship-building and adaptation across great distances and differences, Nabila Hijazi's story of her work with Syrian refugees occurs in close, intimate spaces. Hijazi herself is a Syrian immigrant to the US, yet her chapter narrates events through which she becomes aware that, while she has much in common with the recent Syrian refugee women being welcomed into in her community, her experience of immigration and acculturation also differs from theirs. In other words, because Hijazi willingly immigrated and has had time to adapt to the local culture, her interactions with new refugees creates another kind of intercultural and transnational space. Their cultural identities, languages, and daily practices overlap but also differ, a nuanced complexity that slowly reveals itself through growing dissonance in her community literacy research work. Hijazi describes being deeply engaged with and in service to the Syrian community as she welcomed the incoming refugees, visited their homes and appreciated their hospitality, designed language learning programs, and organized community events. Such investments demonstrate her relational accountability to the women who become her research participants, seeming to pave the way for an easier inquiry process. However, the impetus to adapt still appears: Hijazi describes the "change of heart" emerging as she recognized that the "literacies" she sought to document were determined by Western standards of what counts (i.e., reading and writing) and what does not count (i.e., domestic and child-rearing activities). The West's limited view inaccurately painted her participants as deficient because it erased a whole host of knowledge-based practices not valued in local hierarchies. Her insight led her to revise her study in ways that centered and honored what Syrian refugee women *can* do instead of diminishing them through what they cannot (yet) do.

After adjusting her research focus, Hijazi reflected over how her formal, more detached interview method was a mismatch for the intimate relationships growing out of her community involvement. In "Orienting Framework," she describes how she adapted her process and, as a result, centered her participants. Hijazi realized that storytelling was a better fit for multiple reasons: traditional Syrian culture values the form, it empowered the women to speak for themselves, it emphasized the individualized nature of their experiences, and it evoked complex relations among literacy practices and localized contexts. For example, Hijazi relates a story told to her about the importance of salt when the Syrian war led citizens to be held hostage and deprived of food and medical supplies. That story led her research conversation into reflections over cooking, jam-making, pickling, and other food preservation techniques. On the surface, these conversations may have appeared meandering or even off topic, but Hijazi valued the important literacy information being conveyed *through* the stories. Much more than simply pivoting to revise her interview protocol, Hijazi realigned her project to center her participants and their ways of being in the world. Once her Syrian refugee participants took the priority position, the research process unfolded more gracefully and productively, leading not only to generative outcomes but also to close community relationships that have continued well beyond the project, strengthening reciprocity and relational accountability.

A third story of adaptability as a process of extending a project outward into multiple and complex iterations emerges in "Mingled Threads: A Tapestry of Tales from a Complex Multinational Project" by Rosário Durão, Kyle Mattson, Marta Pacheco Pinto, Joana Moura, Ricardo López-Léon, and Anastasia Parianou. The timeline of the Visualizing Science and Technology across Cultures (VISTAC) project illustrates how it expanded and adapted based on the evolving interests of its large, multinational research team. Durão described the inclusive and flexible design of the project overall, which meant the team of researchers was constantly evolving. That evolution sometimes led to frustration, such as when meetings were unattended or a large array of communication methods had to be assembled, but VISTAC's openness to fluctuation also freed Pinto and Moura to gather their own team of collaborators for the multilingual project. Despite having the survey translated from English into nine different languages, they were surprised to find that participants from some linguistic homes preferred to answer the survey in English instead, as the standard language of scholarly work. Rather than stubbornly clinging to their original survey design, Pinto and Moura reflected over how it could have been adapted into a more effective form.

VISTAC collaborators López-Léon and Parianou also reflected over their revised methods as both used interview and shadowing to study the daily visual literacy activities of engineers. They told stories about adapting their expectations and plans based on the realities of their engineering participants' lives. López-Léon adjusted how he described the project, turned off the audio recorder, and learned to expect multiple conversations with the engineers in order to build trust. Parianou adapted her recruiting plans to cast a wider net and her survey into an in-person questionnaire because the engineers she sought out as her participants were difficult to locate and under other pressures that prevented them from having much time to respond. The austerity measures and socioeconomic stresses that Parianou described are reminders that changes to a project may be in response to factors totally out of the researcher's or practitioner's control. Quality of planning and preparation do not ensure a clear path.

Adaptation may also affect a researcher or practitioner's connection to or role within a project. Mattson's section of their chapter tells the story of him finding his space in the sprawling VISTAC working group. Working then as an assistant professor, he felt pressured to join in but then had to renegotiate his role based on his talents and scholarly knowledge. Mattson had to reflect over and adapt his relation to the project as a whole, which he did out of a sense of accountability. He had to relocate ways he could contribute, even if that meant bowing out of data-gathering opportunities rather than disrupt closer personal relationships in order to participate. Mattson's stories exemplify how complex projects—further complicated by multinational teams and settings—can result in an ongoing repositioning. While Mattson's shifts and decision were a result of team and project design, transnational and intercultural complexities may also trigger repositioning due to any number of factors, such as changing political, economic, or indefinable contexts (as in Longo's project); unexpected responses or outcomes early in the project (e.g., Matheson); changing needs in the participant community; the health and availability of collaborators or participants; or even just growth and evolution due to time spent in a community.

Kathryn Northcut's story reveals how she and her group adapted across disciplinary borders and in response to the COVID-19 pandemic. Joining the team of chemists revolutionized Northcut's entire approach to her scholarly life. She developed a new conception of what "research" looks like, from the deeply collaborative sharing of knowledge to new systems of being visible in the research community (e.g., her ORCHID account and NSF biosketch). Although her chapter traced that evolution in graceful terms, Northcut acknowledged that it was actually very challenging, at times marked by her perceptions of persistent incommensurability, such as when the team reevaluated their approach to changing survey questions on the fly. Northcut's reflections reveal that border crossing can create discomfort at the very core of our professional beliefs and practices. For her, that core was, in part, her careful and conservative approach to following ethical guidelines for research with human subjects. Northcut adapted in a wide variety of ways but met her limits when faced with a fast-and-loose approach to upholding IRB guidelines. Every researcher has such limits, and reading Northcut's story might encourage us to reflect over our own moral principles for being actors in the field.

Despite her thoughtful, agile attitude toward studying ethics in chemistry in both the United States and China, Northcut met up against an insurmountable obstacle in early 2020: the global pandemic caused by the COVID-19. It stopped both her transnational teaching plans and the team's research process—at least temporarily—their tracks. She writes about how their momentum was stymied and then hung in a state of suspended animation as the virus shifted everything. Although we don't yet have the next chapter in Northcut's story, her team likely regrouped, even if on a smaller scale. They might have decided to wait for a vaccine and for travel opportunities to reemerge, or they might have reenvisioned how their project could be adapted to function via video conferencing and other technologies. As Ilunga reminds us, the reaction to facing silences can be to shut down, or it can be to turn toward unlearning and relearning.

Remaining human-centered and relationally accountable may mean loss of control over a project, an anxiety-producing possibility for researchers trained to regulate methods in the name of rigor. Sometimes, it is more ethical to change a project—even if that means the hassle of following up with the IRB—rather than resisting the needs of your participants, collaborators, or the general situation. Running a pilot study can help identify challenges regarding access to collection sites and participants, but even so, researchers must be prepared to tolerate or even welcome uncertainty and to see the unexpected in terms of serendipity rather than derailment. Accountability in transnational and intercultural research requires patience at all phases, from planning and facilitating to writing and reading.

## Principle of Accountable Representation and Reception

Transnational/intercultural research projects require additional care in contextualization and (re)presentation. Teachers, writers, editors, reviewers, and readers share responsibility for setting and maintaining standards.

Some guiding questions follow:

- How will the writers/researchers temper speaking for and about (i.e., Othering) the host country and participants?
- How do disciplinary or organizational norms (e.g., text-based articles) limit representation?
- How does the project approach interpretation and analysis in an inclusive, respectful, and relational manner?
- How can the project outcomes be contextualized in the particular time and place to which they relate?
- How does the project avoid claims of discovery and ownership, of singular and simplified conclusions?
- How will readers react to the inherent limitation of claims in highly contextualized transnational and intercultural research?
- Postdissemination, how will changes to the host site and broader contexts affect ways the project information is used and (re)interpreted? What are the implications for the host stakeholders and participants?

Beyond the actual transnational activity, writing and reading about transnational and intercultural projects entail special care and critical thinking. Writers face ethical questions concerning representation of Others through "texts" (e.g., books, articles, presentations, videos, visuals, conversations, etc.). Also implicated in the dissemination process are layers of audiences who receive these texts, including gatekeepers such as reviewers, editors, and supervisors, as well as various general information providers and consumers in academia, industry, and the public.

Difficulties communicating the complexity of transnational and intercultural research crop up on multiple levels of the reporting process. A common one is the prescriptive force of common academic and workplace genres. In our introduction, Bernadette and I note how the Introduction, Methods, Results, and Discussion (IMRaD) genre diminishes or erases opportunities to discuss the agile and embodied realities of border-crossing work. What we have demonstrated in this collection is that the narratives describing these projects dovetail with project reporting in significant ways. Many of our authors list under "suggested readings" the published studies that complement their narratives. As a field, TPC should not think of

IMRaD and storytelling as distinct and opposing. IMRaD itself is simply a highly disciplined and constrained form of narrative, beginning with the exposition of the literature review, complicated by gaps and issues leading to research questions or hypotheses, directed through a methodical plot, and the ending in a denouement of outcomes. We need a better "both/and" integration of storytelling and project reporting. Versions of that would vary. A few authors are already weaving brief reflections and contextual information into their generally IMRaD driven articles. Our collection makes more substantive space by foregrounding the authors' narratives but leaving space for them to suggest also reading their empirical reports. More permutations of this narrative/expository/persuasive interweaving should be considered and welcomed.

The space for including these stories is at a premium. When a writer is provided with constraints—such as a 5,000–7,000 word limit for an article—they must make decisions about what to include and exclude. Those decisions are compelled by what the reviewers and editors expect. My eventual dissertation, a much more expansive genre than a research article, had a chapter devoted to contextualization, yet two of my three committee members said they didn't want to read it. When revising that dissertation into a book, my editors also said that the contextualizing chapter should be cut. I ended up sneaking in this important information via footnotes and side commentary as an alternative, which worked but felt slightly subversive. While I respected the feedback in both instances, the effective results in both situations were that the transnational and intercultural contextualization was either erased or hidden. Tucking content about gender dynamics, social systems, legal constraints, and more into the nooks and crannies of my work felt exactly like those times I sneaked grated zucchini into my kids' brownies. We as a discipline should prioritize the value of understanding cultural systems if we are to work to address systemic injustices. Scholars must take up more space for contextualization and relevant reflection, and gatekeepers should expect fully developed contextualization as part of demonstrating rigor and ethical representation, particularly in transnational and intercultural projects.

The narrative style of this collection has allowed our authors more room to reflect over their experiences, to write in polyvocal ways, and to explain the complexities of their projects. What it doesn't trace is how the sites of their projects changed over time, how someone visiting these same spaces now might come away with different stories. The collection also cannot fully trace the ways our authors (and we) change over time as we continue learning and evolving, and as we become more aware of histories and processes that have alienated and done violence to others. We learn what not to do but also would benefit from a better framework of what we might do. Our contexts and participants—indeed, we as researchers, practitioners, and teachers ourselves—are constantly evolving. As a result, static textual representations of our places, projects, and selves warp with time. Depending on the host site—and, for example, effects of globalization, political upheaval, or natural forces—change may be rapid and significant, rendering the original research context difficult or impossible to recall. It seems that, perhaps in some projects more than others, we need to account for this. But how?

Across all of our transnational and intercultural projects, we also should carefully attend to the tropes and metaphors we use. For example, it's common to read that research with human subjects is "messy." Such a descriptor may help us find solidarity and encouragement amid complexity and may remind us that neat narratives, prescriptions, and predictions are impossible. However, to say that a transnational and/or intercultural research project is "messy" is also dangerous. Labeling such projects as "messy" risks implying that engaging with representatives of "that Other culture" is what induces the mess, that the Western-trained researcher has been disciplined out of mess and into high standards of order or "rigor." Instead of acknowledging "messiness," researchers should be encouraged by peers, editors, and readers to address complexity, multiplicity, uncertainty, change, adaptation, and serendipity. In fact, we should be encouraged to explore how unfamiliar research contexts reveal the limits of our assumed solid ground and how moving across borders challenges us to find new forms of balance—in terms of ethics, methodologies, methods, and collaborations.

Accountability for quality in our professional activities doesn't stop with practitioners, scholars, and gatekeepers. Our audiences at all levels play a role in upholding shared ethical standards for transnational and intercultural research. Part of that accountability is making space for rich and nuanced contextualization and resisting oversimplification such as Hofstedian-style taxonomies and binaries. Readers must resist coming to intercultural projects comfortable in their traditional disciplinary assumptions about rigor and must work against the impatient impetus to skip over contextualizing exposition and go straight to a project's outcomes. In sum, an effective transnational and intercultural research ethic is an ecology of accountabilities, beginning in our training, proceeding through all the relational aspects of our projects, encompassing our reflections, and shared through our narrated outcomes. From those outcomes then, new and better practices can emerge, and the system of ethical evolution can continue.

## Looking Forward to What's Next

This chapter began with the story of my realization that transnational and cross-cultural research is highly engaging, exciting, compelling, and dangerously fraught. As many of our authors freely admit, we can do our best in preparing, but ultimately, we don't know what we don't know. Depending on your position within our global systems, even participating in intercultural research, teaching, or professional activities may have an undeniable heritage of colonization and injustice. We must look that history in the eye frankly, but we cannot let it stop us from doing better and transformative work. TPC is a field based on border crossing, on evolution, and on envelop pushing. We do our homework, we attend to the quality and ethics of our work, and we challenge ourselves to do better.

Bernadette and I are excited to facilitate sharing these stories with you, but we know that much more work has yet to be done. We need to share more stories, and we need to hear more from TPC practitioners, from projects originating outside of the USA, and from cross-disciplinary and multidisciplinary projects. We need to keep questioning the limitations of our field's dominant genres and celebrating the interesting puzzles presented to us as we work and communicate across complex contexts and differences.

In particular, I'm hopeful our collection might generate momentum for continued conversation about a TPC statement of research ethics for transnational and intercultural projects. What I've proposed here may not be chosen as the best way to proceed, but even if it can serve as a point of generative discussion, it has done its job. In the ongoing effects of globalization and the injustices it perpetuates, a shared disciplinary ethic would provide broad guiding values. Those values, such as being human-centered and relationally accountable, would influence disciplinary norms reinforced through how we teach, learn, perform, and evaluate the work we do. A statement of those values can contribute to the field's continued evolution. That will require a lot of work: developing it will require substantive effort in order to achieve broad disciplinary buy-in. As we are witnessing daily atrocities a national and global levels, a shared statement may seem like an insignificant act. However, even that myriad of injustices results from innumerable small acts and decisions. Small acts can tear down, but small acts also build up and steer us in the directions we want to go.

In addition to stating shared values, a shared research ethic must be accompanied by some kind of framework to assist in operationalizing those values, to help TPC researchers, practitioners, and teachers translate our ideals into on-the-ground, pragmatic activities. Theory is great but must be complimented by opportunities for action. A heuristic or set of guiding questions will never address every situation but can offer a spirit of better practice. It's what I began searching out while I was living in Qatar, it has led me to this moment of what David Levy (2012) called "punctuated equilibrium," yet I want the conversation to continue.

## Acknowledgments

Dr. Tracey Owens Patton's advice to be headstrong and pursue the dignity of my own scholarship inspired and motivated the creation of this collection, and I thank her for her loving and savvy mentorship. Early drafts of the section titled "Toward an Ethic of Transnational and Intercultural Research" were shared with Dr. Rebecca Dingo, Dr. Russell Kirkscey, Dr. Bernadette Longo, and Dr. Rebecca Rickly; their feedback and encouragement have kept the project going. Finally, much of my thinking and motivation for the proposed research ethic were fueled by ongoing debates and pondersome conversations with my colleagues at Texas A&M at Qatar. Thank you to Mysti Rudd, Leslie Seawright, Kelly Wilson, Amy Hodges, Bea Amaya, and Brenda Kent for inspiring me to know more and do better.

## Suggested Readings

Blee, K. M., & Currier, A. (2011). Ethics beyond the IRB: An introductory essay. *Qualitative Sociology*, *34*(3), 401–413.

In this special issue introduction, the authors provide a concise set of perspectives on qualitative ethics and demonstrate, by linking to a variety of studies in social sciences, standard Institutional Review Board (IRB) requirements do not directly address a wide range of situations—at all points of the research and publication process—where ethical entanglements occur.

Gorski, P. C. (2008). Good intentions are not enough: A decolonizing intercultural education. *Intercultural Education, 19*(6), 515–525.

This article unpacks and critiques how current educational endeavors purporting to be "intercultural" actually reinforce stereotypes and hierarchies. Gorski proposes seven shifts in thinking that make progress toward decolonizing thinking and educational processes. All intercultural projects must be in service of creating a more socially just world.

Katz, S. (1992). The ethic of expediency: Classical rhetoric, technology, and the Holocaust. *College English, 54*(3), 255–275.

Katz's foundational article horrifyingly illustrates the potential dangers of defaulting to a limited practical approach to communication situations and processes. Although this article is focused on technological ethos, its warning extends to the oversimplification of culture and difference into limited, prescriptive taxonomies.

Ratcliffe, K. (2005). *Rhetorical listening: Identification, gender, whiteness.* Studies in Rhetorics and Feminisms. Southern Illinois University Press.

Ratcliffe proposes a theory and process for deep, active listening that attends particularly to tropes that inform cultural logics. Her method is based on "standing under" (as an effort toward understanding), accountability and interdependency, and acknowledgment of both commonalities and differences.

Rudd, L. (2018). "It makes us even angrier than we already are": Listening rhetorically to students' responses to an honor code imported to a transnational university in the Middle East. *Journal of Global Literacies, Technologies, and Emerging Pedagogies, 4*(3), 655–674.

The author describes a research project set in an international branch campus and the struggles that emerged directly in relation to colonizing power flows. Rudd tells the story of how the project evolved as she and her colleagues turned to rhetorical listening, adjusted their own frameworks, and learned to value the local voices over the imperatives of Western educational tradition.

Shope, J. H. (2006). "You can't cross a river without getting wet": A feminist standpoint on the dilemmas of cross-cultural research. *Qualitative Inquiry*, *12*(1), 163–184.

This article was one of the first I could locate that frankly described how her project grappled with the challenges of ethically working across cultures. Shope applies principles of feminist theory to sit with, rather than neatly resolve, tensions emerging from multiple aspects of her project.

Small, N. (in press). *A rhetoric of becoming: USAmerican women in Qatar.* Studies in Rhetorics and Feminisms. Parlor Press.

This monograph considers how the stories expatriate women in Qatar told each other and folks back home in the US created a shared narrative lifeworld rooted in paradoxical positionings. In the final chapter, I provide the emergent feminist practice of "micropraxis," small rhetorical moves intended to purposefully address injustices. Micropraxis as a theory and practice can also inform better research design.

Walton, R. (2016). Supporting human dignity and human rights: A call to adopt the first principle of human-centered design. *Journal of Technical Writing and Communication*, *46*(4), 402–426.

Walton establishes the valuable connection between technical and professional communication and human-centered design principles. Those principles, based in rights and dignity while facilitated through communication, can lead TPC scholars and professionals through the complex process of deciding who (among competing communities and stakeholders) is centered.

## Discussion Questions

1. In the opening story about the female majlis presentation, Matthew and Dr. Qadha are more critical of the research talk because of their positioning in relation to the local culture. What are some strategies that cultural outsiders—like the author of this chapter—might use to

be more informed and thoughtful consumers of other people's transnational and intercultural research?

2. The author described how she and her colleagues formed a discussion group to explore the ethics of their transnational projects and that the discussion group often felt like it was going in circles and reaching dead ends. If questions of ethics and transnational research are so challenging, then would we be wise to limit our research only to those people and activities in our own "home" cultures?
3. In another story, the author described a visiting research team from the US that wanted to study Qatari women. When the author asked them "What's in it for your Qatari participants?" the response was "We can help them see how education is improving their lives." What would you have said or done in reply to their response?
4. This chapter hints at several tensions regarding transnational research. What tensions do you notice or relate to?
5. What principles and/or reflective questions would you add to the lists under "Toward an Ethic of Transnational and Intercultural Research"?
6. Developing a disciplinary or professional shared ethic is a collaborative process. How would you imagine that process unfolding? Who would lead it? Who would participate? How would it be facilitated?

## Author Biography

**Dr. Nancy Small** is an assistant professor of English and director of First Year Writing at the University of Wyoming. She studies everyday women's storytelling as a form of shared sense-making and a process of building shared rhetorical lifeworlds. Her work has appeared in *Peitho: Journal of the Coalition of Feminist Scholars in the History of Rhetoric & Composition*; *Kairos: A Journal of Rhetoric, Technology, and Pedagogy*; and the *Journal of Technical Writing and Communication.* Her study on the lives of white USAmerican women living and working in Qatar, is titled *A Rhetoric of Becoming: USAmerican Women in Qatar* (in press).

# References

Adichie, C. N. (2009, Oct. 7). *The danger of a single story.* [Video]. TED. https://www.youtube.com/watch?v=D9Ihs241zeg

Agboka, G. Y. (2014). Decolonial methodologies: Social justice perspectives in intercultural technical communication research. *Journal of Technical Writing and Communication*, *44*(3), 297–327.

Alcoff, L. (1991). The problem of speaking for others. *Cultural Critique*, *20*, 5–32.

Álvarez, Steven. (2017). "Latinx and Latin American Community Literacy Practices en Confianza." *Composition Studies*, *45*(2), 219–221.

Anzaldúa, G. (1999). *Borderlands/la frontera*. Aunt Lute.

Archibald, J. A., Lee-Morgan, J., & De Santolo, J. (Eds.). (2019). *Decolonizing research: Indigenous storywork as methodology*. Zed Books.

Aspen Institute. (August 2011). "Agent Orange/Dioxin History." Retrieved: July 6, 2020. https://www.aspeninstitute.org/programs/agent-orange-in-vietnam-program/agent-orangedioxin-history/

Bankert, E., Gordon, B., Hurley, E., & Shriver, S. (2020). *Institutional Review Board: Management and function* (3rd ed.). Jones and Bartlett Publishers International.

Bazerman, C., Paradis, J. G., & Paradis, J. (Eds.). (1991). *Textual dynamics of the professions: Historical and contemporary studies of writing in professional communities*. University of Wisconsin Press.

Beauchamp, T. L. (2011). Informed consent: Its history, meaning, and present challenges. *Cambridge Quarterly of Healthcare Ethics*, *20*, 515–523. http://doi.org/10.1017/S09631801110000259

Benjamin, W. (1969). *Illuminations*. (H. Zohn, Trans.). Schocken Books.

Bennett, J., Muholi, Z., & Pereira, C. (2012). *Jacketed women: Qualitative research methodologies on sexualities and gender in Africa*. University of Cape Town Press.

Blee, K. M., & Currier, A. (2011). Ethics beyond the IRB: An introductory essay. *Qualitative Sociology*, *34*(3), 401–413.

Booth, W. C. (2004). *The Rhetoric of rhetoric: The quest for effective communication*. Blackwell.

Bowles, D. (2020). Cummins' non-Mexican crap [blog post]. *Medium*. Retrieved from https://medium.com/@davidbowles/non-mexican-crap-ff3b48a873b5

Bosley, D. (Ed.). (2001). *Global contexts: Case studies in international technical communication*. Allyn & Bacon.

Brooks, L. T. (2008). *The common pot: The recovery of Native space in the Northeast*. University of Minnesota Press.

Buchanan, R. (2001). Human dignity and human rights: Thoughts on the principles of human-centered design. *Design Issues*, *17*(3), 35–39.

Cain, M. A. (1994). Mentoring as identity exchange: conflicts and connections. *Feminist Teacher*, *8*(3), 12–118.

Carbaugh, D. (Ed.). (1990) *Cultural communication and intercultural contact*. Lawrence Erlbaum.

Charmaz, K. (2006). *Constructing grounded theory: A practical guide through qualitative analysis*. Sage Publications.

Chaudhuri, N., Katz, S. J., & Perry, M. E. (Eds.). (2010). *Contesting archives: Finding women in the sources*. University of Illinois Press.

Cruikshank, J. (1998). *The social life of stories: Narrative and knowledge in the Yukon Territory*. University of Nebraska Press.

Cummins, J. (2020). *American dirt*. Flatiron Books.

De Certeau, M. (1988). *The writing of history*. Columbia University Press.

Ding, H., & Savage, G. (2013). Guest editors' introduction: New directions in intercultural professional communication [Special issue]. *Technical Communication Quarterly*, *22*(1), 1–9.

Eble, M. F. (2008). Reflections on mentoring. In L. G. Eble & M. F. Eble (Eds.). *Stories of mentoring: Theory & praxis* (pp. 306–312). Parlor Press.

Emanuel, E., Gadsden, A., & Moore, S. (April 19, 2019). How the U.S. surrendered to China on scientific research. *The Wall Street Journal*, C3.

EMTV Online. (2017, May 23). *Opening of new classroom in remote part of Nipa Kutubu* [Video]. YouTube. https://youtu.be/9LBVg6hLShE

Erdrich, L. (2003). *Books and islands in Ojibwe country*. Harper Collins.

Foss, K. A, & Foss S. K. (1994). Personal experience as evidence in feminist scholarship. *Western Journal of Communication*, *58*(1), 39–43.

Fraiberg, S. (2013). Reassembling technical communication: A framework for studying multilingual and multimodal practices in global contexts. *Technical Communication Quarterly*, *22*(1), 10–27.

Freire, P. (1971). *Pedagogy of the oppressed*. New York: Herder and Herder.

Glenn, C. (1997). *Rhetoric retold: Regendering the tradition from antiquity through the Renaissance*. Southern Illinois University Press, 1997.

Glenn, C., & Ratcliffe, K. (2011). *Silence and listening as rhetorical arts*. Southern Illinois University Press.

Gonzales, L. (2018). *Sites of translation: What multilinguals can teach us about digital writing and rhetoric*. University of Michigan Press.

Gorski, P. C. (2008). Good intentions are not enough: A decolonizing intercultural education. *Intercultural Education, 19*(6), 515–525.

Groznya, E. (2013). Lost in translation. In H. Yu & G. Savage (Eds.). *Negotiating cultural encounters: Narrating intercultural engineering and technical communication* (pp. 81–101). IEEE.

Haas, A. M. (2007). Wampum as hypertext: An American Indian intellectual tradition of multimedia theory and practice. *Studies in American Indian Literatures, 19*(4), 77–100.

Haas, A. M., & Eble, M. F. (Eds.). (2018). *Key theoretical frameworks: Teaching technical communication in the twenty-first century*. University of Colorado Press.

Harris, R. A. (Ed.). (2005). *Rhetoric and incommensurability*. Parlor Press.

Haswell, R. H. (2005). NCTE/CCCC's recent war on scholarship. *Written Communication, 22*(2), 198–223.

Hofstede, G. (2001). *Culture's consequences: Comparing values, behaviors, institutions and organizations across nations* (2nd ed.). Sage.

Hofstede, G. H., Hofstede, G. J., & Minkov, M. (2010). *Cultures and organizations: Software of the mind*. McGraw-Hill.

Holborow, M. (1999). *The politics of English*. Sage.

Hunsinger, R. P. (2006). Culture and cultural identity in intercultural technical communication. *Technical Communication Quarterly, 15*(1), 31–48.

Israel, B. A. (2001). Community-based participatory research: Policy recommendation for promoting a partnership approach in health research. *Education for Health, 14*(2), 182–197.

Itchuaqiyaq, C. U., & Matheson, B. (2021). Decolonizing decoloniality: Considering the (mis)use of decolonial frameworks in TPC scholarship. *Communication Design Quarterly Review, 9*(1), 20–31.

Kalichman, M. (2014). Rescuing responsible conduct of research (RCR) education. *Accountability in Research, 21*(1), 68–83.

Kamwangamalu, N. M. (2003). Globalization of English, and language maintenance and shift in South Africa. *International Journal of the Sociology of Language, 164*, 65–81.

Katz, S. (1992).The ethic of expediency: Classical rhetoric, technology, and the Holocaust. *College English, 54*(3), 255–275.

Keller, E. J. (2018). *Rhetorical strategies for professional development: Investment mentoring in classrooms and workplaces*. Routledge.

Kimmerer, R. W. (2013). *Braiding sweetgrass: Indigenous wisdom, scientific knowledge, and the teachings of plants*. Milkweed.

King, L., Gubele, R., & Anderson, J. R. (Eds.). (2015). *Survivance, sovereignty, and story: Teaching American Indian rhetorics*. University of Colorado Press.

King, T. (2003). *The truth about stories: A native narrative.* House of Anansi.

Latour, B. (1987). *Science in action: How to follow scientists and engineers through society.* Harvard University Press.

Levy, D. M. (2012). *Scrolling forward: Making sense of documents in the digital age.* Simon and Schuster.

Longo, B. (1998). An approach for applying cultural study theory to technical writing research. *Technical Communication Quarterly, 7*(1), 53–73.

Longo, B. (2000). *Spurious coin: A history of science, management, and technical writing.* State University of New York Press.

Longo, B. (2014). R U there? Cell phones, participatory design, and intercultural dialogue. *IEEE Transactions on Professional Communication, 57*(3), 204–215.

Longo, B. (2014). Using social media for collective knowledge-making: Technical communication between the global north and south. *Technical Communication Quarterly, 23*, 22–34.

Longo, B. (2018, June 28). Humanizing computer history. *Humanities for STEM Symposium Proceedings*, New York University, April 6–7, 2018. https://osf.io/79ra5/

Longo, B., & Ilunga, Y. Y. (2012). *Trust* [Video]. TEDxNJIT. Retrieved from https://www.youtube.com/watch?v=Q3_nqibpkjU

Louhiala-Salminen, L., & Kankaanranta, A. (2012). Language as an issue in international internal communication: English or local language? If English, what English? *Public Relations Review, 38*(2), 262–269.

Matheson, B., & Petersen, E. J. (2020). Engaging U.S. students in culturally aware content creation and interactive technology design through service learning. *IEEE Transactions on Professional Communication, 63*(2), 188–200.

Maylath, B., Vandepitte, S., Minacori, P., Isohella, S., Mousten, B., & Humbley, J. (2013). Managing complexity: A technical communication translation case study in multilateral international collaboration. *Technical Communication Quarterly, 22*(1), 168–184.

McSweeney, B. (2002). Hofstede's model of national cultural differences and their consequences: A triumph of faith—a failure of analysis. *Human Relations, 55*(1), 89–118.

Meyers, S. V. (2014). *Del otro lado: Literacy and migration across the U.S.-Mexico Border.* Southern Illinois University Press.

Miller, C. R. (1979). A humanistic rationale for technical writing. *College English, 40*(6), 610–617.

Moore, K. R., Meloncon, L., & Sullivan, P. (2017). Mentoring women in technical communication. In Kirsti Cole & Holly Hassel (Eds.). *Surviving sexism in academia: Strategies for feminist leadership* (pp. 233–240). Routledge.

Moses, M. G., & Katz, S. B. (2006). The phantom machine: The invisible ideology of email (A cultural critique). In J. B. Scott, B. Longo, & K. V. Wills

(Eds.). *Critical power tools: Technical communication and cultural studies* (pp. 71–105). State University of New York Press.

Noe-Bustamante, L., Flores, A. (2019). *Facts on Latinos in the U.S.* https://www.pewresearch.org/hispanic/fact-sheet/latinos-in-the-u-s-fact-sheet/

Nussbaum, M. (1997). *Cultivating humanity: A classical defense of reform in liberal education.* Harvard University Press.

Ornelas, I. J., Yaminis, T. J., Ruiz, R. A. (2020). The health of undocumented Latinx immigrants: What we know and future directions. *Annual Review of Public Health, 41*, 289–308.

Petersen, E. J. (2017). Feminist historiography as methodology: The absence of international perspectives. *connexions: international professional communication journal, 5*(2), 1–38.

Petersen, E. J., & Matheson, B. (2017, August). Following the research internationally: What we learned about communication design and ethics in India. In *Proceedings of the 35th ACM International Conference on the Design of Communication* (pp. 1–6).

Pigozzi, L. (2018). Negotiating informed consent: *Bueno aconsjar, major remdiar* (it is good to give advice, but it is better to solve the problem). In L. Meloncon & J. B. Scott (Eds.). *Methodologies for the rhetoric of health & medicine* (pp. 195–213). Routledge.

Pigozzi, L. (2020). *Caring for and understanding Latinx patients in health care settings.* Jessica Kinsley Publishers.

Powell, M. D. (2004). Down by the river, or how Susan La Flesche Picotte can teach us about alliance as a practice of survivance. *College English, 67*(1), 38–60.

Ramsey, A. E., Sharer, W. B., L'Eplattenier, B., & Mastrangelo, L. (Eds.). (2009). *Working in the archives: Practical research methods for rhetoric and composition.* Southern Illinois University Press.

Ratcliffe, K. (2005). *Rhetorical listening: Identification, gender, whiteness.* Southern Illinois University Press.

Riley-Mukavetz, A. (2020). Developing a relational scholarly practice: snakes, dreams, and grandmothers. *College composition and communication, 71*(4), 545–565.

Rönkkö, K., Hellman, M., Kilander, B., & Dittrich, Y. (2004). Personas is not applicable: Local remedies interpreted in a wider context. *Proceedings of the Participatory Design Conference 2004*, 112–20.

Royster, J. J. (1996). When the first voice you hear is not your own. *College Composition and Communication, 47*(1), 29–40.

Rudd, L. (2018). "It makes us even angrier than we already are": Listening rhetorically to students' responses to an honor code imported to a transnational university in the Middle East. *Journal of Global Literacies, Technologies, and Emerging Pedagogies, 4*(3), 655–674.

Samuelson, B. L., & Freedman, S. W. (2010). Language policy, multilingual education, and power in Rwanda. *Language Policy*, *9*(3), 191–215.

Savage, G., & Sullivan, D. (Eds.). (2001). *Writing a professional life: Stories of technical communicators on and off the job.* Allyn & Bacon.

Scott, J. B., & Longo, B. (2006). Guest editors' introduction: Making the cultural turn [Special issue]. *Technical Communication Quarterly*, *15*(1), 3–7.

Shope, J. H. (2006). "You can't cross a river without getting wet": A feminist standpoint on the dilemmas of cross-cultural research. *Qualitative Inquiry*, *12*(1), 163–184.

Sides, C. H. (Ed.). (1989). *Technical and business communication: Bibliographic essays for teachers and corporate trainers*. National Council of Teachers of English.

Slack, J. D., Miller, D. J., & Doak, J. (1993). The technical communicator as author: Meaning, power, authority. *Journal of Business and Technical Communication*, *7*(1), 12–36.

Small, N. (2017). (Re)Kindle: On the value of storytelling to technical communication. *Journal of Technical Writing and Communication*, *47*(2), 234–253.

Small, N. (2022). Localize, adapt, reflect: A review of recent research in transnational and intercultural TPC. In L. Melonçon & J. Schreiber (Eds.). *Assembling critical components: A framework for sustaining technical and professional communication* (pp. 269–295). Colorado State University Press.

Small, N. (in press). *A rhetoric of becoming: USAmerican women in Qatar.* Parlor Press.

Smith, L. T. (1999). *Decolonizing methodologies: Research and Indigenous peoples.* Zed Books.

Spivak, G. C. (2003). Can the subaltern speak? *Die Philosophin*, *14*(27), 42–58.

Spivak, G. C. (1999). *A critique of postcolonial reason.* Harvard University Press.

Tercedor-Sánchez, M. I., & Abadía-Molina, F. (2005). The role of images in the translation of technical and scientific texts. *Meta*, *50*(4).

Tuck, E., & Yang, K. W. (2012). Decolonization is not a metaphor. *Decolonization: Indigeneity, Education & Society*, *1*(1), 1–40.

Thatcher, B. (2001). Issues of validity in intercultural professional communication research. *Journal of Business and Technical Communication*, *15*(4), 458–489.

Thatcher, B., & St. Amant, K. S. (Eds.). (2011). *Teaching intercultural rhetoric and technical communication: Theories, curriculum, pedagogies, and practices.* Baywood.

Thorp, L. (2006). *The pull of the earth: Participatory ethnography in the school garden* (Vol. 7). Rowman Altamira.

United States Senate Committee on Homeland Security and Governmental Affairs. (2019, Nov. 18). *Threats to the U.S. research enterprise: China's talent recruitment plans.* https://www.hsgac.senate.gov/imo/media/doc/2019-11-18%20PSI%20Staff%20Report%20-%20China's%20Talent%20Recruitment%20Plans.pdf

Walton, R. (2016). Supporting human dignity and human rights: A call to adopt the first principle of human-centered design. *Journal of Technical Writing and Communication*, *46*(4), 402–426.

Walton, R., Moore, K., & Jones, N. (2019). *Technical communication after the social justice turn: Building coalitions for action*. Routledge.

Walton, R., Zraly, M., & Mugengana, J. P. (2015). Values and validity: Navigating messiness in a community-based research project in Rwanda. *Technical Communication Quarterly*, *24*(1), 45–69.

Wardlow, H. (2006). *Wayward women: Sexuality and agency in New Guinea society*. University of California Press.

Williams, M. & Pimentel, O. (Eds.). (2014). *Communicating race, ethnicity, and identity in technical communication*. Baywood.

Williams, R. (1976/2015). *Keywords: A vocabulary of culture and society*. Oxford University Press.

Wilson, S. S. (2008). *Research is ceremony: Indigenous research methods*. Fernwood Publishing.

Windchief, S., & San Pedro, T. (Eds.). (2019). *Applying Indigenous research methods: Storying with peoples and communities*. Routledge.

Yu, H., & Savage, G. (Eds.). (2013). *Negotiating cultural encounters: Narrating intercultural engineering and technical communication*. John Wiley & Sons.

www.ingramcontent.com/pod-product-compliance
Lightning Source LLC
LaVergne TN
LVHW050151080826
844660LV00002B/171

* 9 7 8 1 4 3 8 4 8 9 0 3 2 *